BUILDING FAILURES
Recovering the Cost

BUILDING FAILURES
Recovering the Cost

Alan Crocker

RIBA, DipArch, ACIArb

Foreword by
The Right Honourable The Lord Scarman
OBE, PC, DCL, LLD

BSP PROFESSIONAL BOOKS
OXFORD LONDON EDINBURGH
BOSTON MELBOURNE

First published 1990

British Library
Cataloguing in Publication Data
Crocker, Alan
  Building failures: recovering the cost.
  1. Great Britain. Buildings. Construction.
  Law
  I. Title
  344.1037869

ISBN 0-632-02515-8

BSP Professional Books
A division of Blackwell Scientific
  Publications Ltd
Editorial Offices:
Osney Mead, Oxford OX2 0EL
  (Orders: Tel. 0865 240201)
25 John Street, London WC1N 2BL
23 Ainslie Place, Edinburgh EH3 6AJ
3 Cambridge Center, Suite 208, Cambridge
  MA 02142, USA
54 University Street, Carlton, Victoria 3053,
  Australia

Set by DP Photosetting, Aylesbury, Bucks
Printed and bound in Great Britain by
Billing & Sons, Worcester

For Jo, with love and thanks for offering encouragement and constructive criticism in just the right proportions and for her uncomplaining acceptance of word-processor widowhood.

Contents

Foreword
by
The Right Honourable the Lord Scarman
OBE, PC, DCL, LLD

Most of us are at some time or another 'building owners'. A building owner is one who owns or rents land or a building and decides either to build something new on the land or to extend or repair what is already there.

In any of these situations you, the owner, may have to rely on the advice of an architect, a surveyor, or a civil engineer and will almost certainly have to engage a builder. With each you will have a contract upon which you will have to rely if any of your expert advisers or your builder should let you down. But what is involved in going to law or arbitration to recover from any of your advisers or from your builder compensation for loss arising from their work under contract with you? And how do you prove that your loss was caused by his (or her) breach of contract? All you know is that the building is defective in certain respects: but by what or by whom was the defect caused? You may be able to make a shrewd guess, but can you *prove* it?

Building litigation is a complex and expensive business. Alan Crocker knows all about it, having learned it the hard way by years of experience as a technical adviser and expert witness. His book, for which we must be truly thankful, is written not in jargon, but in good ordinary English. It is a fascinating journey that the reader takes with the author, as he charts the progress of a building case from the discovery of building failure to (we sincerely hope) the successful conclusion of a judgment or award of damages to the claimant together with his legal costs.

It is, of course, a journey which can end in defeat, frustration, and heavy financial loss. Our author charts the dangers of the passage. You need a really competent and experienced guide. The real lesson of this book is to choose carefully your expert adviser and your lawyer. Alan Crocker tells you how to go about the task of choosing. And today solicitors are allowed to 'advertise' their specialities. Building cases require all the expert help you can get: they are not suitable territory for a 'do-it-yourself' expedition. Go into battle with the best equipped and trained troops you can muster.

Mr. Crocker's book fills a gap in the knowledge of most of us and should be on the bookshelf at home and in the office as an indispensable first aid, ready to hand when trouble comes. But it is more than a safety net: it is also a good 'read', telling a fascinating story of little known but all too frequent litigation in a specialist field of importance to all of us.

Scarman
4 May 1990

Preface

I have been involved for many years in dealing with the consequences of building defects, mostly, I am relieved to say, not of my making. This involvement has dealt with both technical and legal aspects. With regard to the technical, my work has included the diagnosis of causes of failure and the design and implementation of both temporary and permanent remedial work. The legal side has required liaison with lawyers and technical professionals in disciplines other than my own. It has required me to make assessments of responsibility for failure, to prepare evidence and opinions in the form of reports and to confirm them in the witness box.

During the course of many different cases a number of things have become apparent which, to me, indicated the need for this book. Without in any way denigrating the knowledge and experience of either my fellow designers or the lawyers with whom I have come into contact, or my clients, I make the following representations.

There are highly skilled motor mechanics who, after a working lifetime servicing Fords, would have considerable problems faced with diagnosing an esoteric operating fault in a Ferrari. There are chefs who, after years of producing exotic desserts, would be hard put to make a worthwhile soup. There are respected and highly skilled orthopaedic surgeons who would run a four minute mile from a brain operation. Likewise, there are solicitors highly competent in many fields, but with no knowledge of building litigation, and excellent architects who would not know how to set about finding the cause of a defect.

In a different way, there are solicitors who know much about building disputes but not very much about how architects and other designers work, nor how experts set about diagnosing the causes of building defects. Perhaps even more common, there are the technical experts who know nothing about the law, or, worse, know a little.

Consider also their clients. Those with defective buildings, apart from some with large property portfolios, who possibly will know nothing about the law or building. Those alleged to be responsible for the defects will probably know not very much about the law and perhaps (not always) not quite enough about building.

This book has, therefore, been written to set down as logically as possible the events that occur from the onset of symptoms of failure to the ultimate resolution of the problems in one form or another. It is intended to explain to all of the parties to a dispute, and to their advisers, the role of everyone as well as the events that take place and, as far as practicable, the sequence in which they do so. It is also hoped that it will explain the importance of getting suitable specialist advice at an early stage, and how to get it, and of each person having an awareness of the role of the others and the need for teamwork. In consequence, it is inevitable that some readers will feel that in the discussion of some matters they are being talked down to. Perhaps however, even the experienced will find some useful enlightenment with regard to facets of another's work that they had not previously appreciated. It may also be of interest and help to the many people who suffer the consequences of building defects without even realising that it may be possible to recover a substantial part of their expenditure.

As far as the law is concerned, conscious of my own dangerously limited knowledge of the subject, I have kept mention of legal matters to the minimum consistent with pointing the reader in the right direction.

I thank John Parris, and his publishers, BSP Professional Books, for permission to refer to his book, *Arbitration: Principles and Practice*; and the Building Research Establishment for permission to reproduce, as background for the jacket, an illustration from Eldridge's classic book on building defects. This illustration is a Crown copyright Building Research Establishment photograph.

Finally, this seems a fitting opportunity to recall from my schooldays the memory of the deputy head, 'Sandy' Southerst, and the head of English, 'Soapy' Sleight. One persuaded me to study architecture, the other wanted me to become a writer. I leave it to others to decide whether either of them would now be pleased.

Alan Crocker
1990

Acknowledgements

I have had many discussions concerning this book with my partner, Elizabeth Tooth, all of them to my great benefit.

I am particularly grateful to Roger Button, Head of the Building and Engineering Department of Jaques and Lewis, solicitors. He has generously made available to me his time, knowledge and expertise and as a result saved me from a number of errors and inaccuracies. However, the final decisions on content have been mine, as is responsibility for the end product.

Introduction

While accurate figures are hard to come by, it has been estimated that the annual cost of building maintenance and repair in the UK is currently in the region of £9000 million. Of this sum, about 40% is spent on buildings erected since 1960, and of that roughly half is the consequence of unnecessary failure rather than reasonable and legitimate maintenance. At a very much rounded down and conservative estimate this means that over £1000 million at present day values, is spent annually on correcting defects which are the direct consequence of premature failure.

In 1987 3600 writs were issued covering building disputes, an increase of 40% since 1983. According to RIBA indemnity research, in 1981, for every ten architect's professional indemnity insurance policies (policies covering successful negligence claims against the architect) in force there was one claim annually. By 1987 this had risen to seven claims for every ten policies. The majority of the writs were in consequence of some sort of building defect or failure, and the indemnity policy claims the result of alleged designers' faults which caused or contributed to failure.

The experience of the author, borne out by Building Research Establishment statistics, is that most failures are caused by design and/or workmanship errors, a few are attributable to faulty materials or errors of procedure, and even fewer to poor maintenance.

A building owner faced with substantial repair bills and financial loss due to disruption of his business, loss of rent or other causes will try, understandably, to recover the cost from someone. However, in the greatest number of instances the type and cause of damage he has suffered will not be covered by his own insurance.

To assist him, he can obtain various forms of practical advice and help from specialist legal and technical advisers.

They, while knowing their own job, do not always fully appreciate the work done and the difficulties experienced by their counterparts in other disciplines. More commonly, the building owner, and those he believes responsible for his problems, have an inadequate notion of how to set about obtaining advice, and what happens when that advice is transformed into action.

The purpose of this book is to explain the processes involved from the first symptoms of trouble to the point at which the owner either agrees a settlement or an award is made by a court or arbitrator. First, it is as well to be quite clear as to what constitutes the sort of problem that results in a dispute.

What is building failure?

In many respects buildings are similar to people. Most have a frame (the skeleton), cladding (the skin) and services. The last of these include the machinery to deal with the intake of fuel, water supply, means of heating and ventilation, disposal of waste, and others not too far removed from some of the functions of the human body.

People deteriorate as part of the natural processes of ageing, and so do buildings. People also deteriorate, either temporarily or permanently, as a consequence of some malfunction, and, again, so do buildings.

It is failure resulting from malfunction, rather than old age, that is dealt with in this book. It may therefore be described as premature failure resulting from errors of design, workmanship, maintenance or the use of faulty materials. The term design, incidentally, means the total design process which includes the specification of either materials and workmanship or, alternatively, performance. It includes the design input of everyone in the building team with an appropriate design responsibility.

Symptoms of failure and damage

Returning to the analogy of the human body, illness is normally, although not inevitably in the early stages, recognised after the appearance of symptoms. Equally, in the event of illness, some damage will occur, either temporary or permanent. In some cases illness may respond to treatment by drugs, in others surgery may be necessary.

Damage

Similarities exist with buildings. Symptoms of failure in buildings vary enormously and may herald considerable variations in the severity of damage. Later in the book the importance of trying to define accurately damage in the particular context of building defects will be discussed: it is not always as simple as it would first appear.

The shorter Oxford English Dictionary describes damage as 'Loss or detriment caused by hurt or injury affecting estate, condition, or circumstances. Injury, harm. A disadvantage, a misfortune.' In terms of recovery of the cost of failure, it is the question of loss which is of paramount importance, and, consequentially, establishing that loss, and particularly financial loss, has occurred.

Causes of failure

It is of course also essential to establish the cause or causes of failure and resultant damage, which will, when considering rectification, profoundly affect the way forward. Apart from the detailed technical reasons for failure, causes fall into the following basic categories:

(1) Natural phenomena such as storms, resulting in damage from floods, exceptionally high winds, lightning. Earthquakes.
(2) Design errors.
(3) Workmanship errors and faulty materials.
(4) Procedural errors.
(5) Failure to maintain properly.
(6) Abuse or misuse of the building.

Responsibilities

At first sight it would seem that (1) and (5) above are in no way the responsibility of either designers or builders. This may often be so, but not necessarily. The third of course may be entirely the responsibility of contractor or sub-contractor, or may also involve others with a duty to inspect works in progress. It can also, in many cases, be the duty of a contractor to warn of the existence of design errors.

The first category may or may not be the concern of either designers or builders, depending on the extent to which the natural phenomenon should have been anticipated and designed for. The fourth could involve designers if it can be established that owners would not have been fully aware of maintenance requirements without being given explicit advice, or if maintenance requirements are excessive.

It will be seen therefore that damage resulting from natural causes may normally be regarded by insurers as something that could, and should, have been guarded against by either designer or owner, while damage resulting from lack of maintenance may well be considered by a building owner to have been the consequence of a failure on the part of designers to give specialist advice. It should, incidentally, be noted that designers may well include manufacturers of specialist materials, components, or equipment.

The incidence and severity of failure

It is regrettable that, at the end of the 1980s, building failure in Britain is widespread mostly in comparatively new buildings of many types, large and small.

The majority of defects are to be found in buildings completed since the start of the post second world war building boom. At present the amount of remedial work needed to correct the consequences of premature failure is enormous, and a large number of cases of failure are being dealt with by litigation and arbitration.

The cost of failure to individual buildings

As far as individual cases of failure are concerned, it is obvious that costs will vary enormously depending on the size of project, the nature and extent of the failure, the occurrence of consequential damage and loss, and the procedures involved in putting things right.

What must be borne in mind, however, is that by the time fees, loss of profits, rents, etc., are taken into account, the total cost of rectification is usually much higher than the cost of the actual remedial work.

Who is involved?

In some cases, a building owner suffering damage to his building, and perhaps consequential loss due to loss of profits, damage to furnishings, machinery, stock or other items (quite apart from the possibility of injury to persons) may be fortunate enough to settle his claim directly with whoever is responsible, be it architect, engineer or contractor.

More often than not however, designers, contractors and suppliers will each blame someone else, and may suggest that the problems are due to some abuse by the occupants or failure to maintain the premises properly. When this happens, each party is likely to become involved with their own solicitors, barristers, technical experts and insurers, quite apart from each other and, ultimately, a judge or an arbitrator. All this apart from any number of witnesses of fact.

In the following pages it is hoped to show what all those involved do, as well as how, why and when.

Chapter 1

Types of failure and the consequences

The introduction defined failure and described causes stemming from design, workmanship, materials, procedural deficiencies or poor maintenance. In other words, failures that would not have occurred, or would not have occurred prematurely, without the influence of such deficiencies. The following are examples of each:

- Design (which includes specifications of performance or of workmanship and materials): using roofing tiles at below the recommended pitch, resulting in excessive ingress of water.
- Workmanship: failing to use the correct number and spacing of wall ties in a cavity brick wall, with a resulting loss of stability.
- Materials: a batch of structural steel which appears sound but has evaded quality control checks, and is in fact sub-standard.
- Procedures: working out of sequence making it virtually impossible for a specialist sub-contractor to carry out all his work correctly.

It is stressed that it is common for failures to be ascribed to faulty materials when in fact there is nothing wrong with the materials but much wrong with the manner of their use. This may result from either design or workmanship error. In fact, failure due to faulty materials is statistically shown to be the least common cause after poor maintenance.

Faulty workmanship, and very occasionally faulty materials, can sometimes be of a nature that should be seen, in part at least, by inspection of work in progress. Such inspection may be by designer, clerk of works, contractor's or sub-contractor's supervisory staff or by a building control officer. A later chapter will consider the implications of the failure by one or more persons to see what could or should have reasonably been seen.

Failures due to design faults

As far as design is concerned, a number of people claim that it is not usually conceptual design which is at fault, but the detailing. In fact, a competent designer will have at least the principles of good detailing in his mind right from the start of the design process.

It is true that a good basic design may be partly upset by poor and unchecked detailing during the production drawing stage. It is equally certain that many buildings which fail prematurely do so because knowledge of good detailing principles and choice of materials were not possessed or the detailing was not considered worthy of thought in the early design stage. Another frequent reason is innovative design undertaken without adequate research.

Failures due to poor workmanship

With regard to workmanship, there may be deficiencies in the labour of both main and sub-contractors. These may be the consequence of lack of skill, lack of care and interest, or a lack of knowledge of the importance of special care in the execution of some vital piece of work. This last however could equally well arise because the designer has failed to stress the unusual nature of some design feature, or the contractor has failed to pass on some vital piece of information to operatives. Indeed, it is not unknown for a designer to visit a site and discover that tradesmen do not have a copy of the specification to hand and have never even seen one.

Faulty materials

Failures of materials in the true sense are comparatively rare, although some of the materials in use today are innovative and launched in the market place without adequate testing, or with the virtues overemphasised and the possible failings played down.

Procedures as failure inducers

Procedures are also relatively uncommon as sources of failure, but are sometimes responsible. This is particularly so when delays occur in the delivery of materials or components, late changes in sub-contractor or supplier are made, or when instructions are given to change design details or materials and in consequence unauthorised extra expenditure is incurred, or other problems result. Frequently the problems arise due to a failure to check the effect of change on other related details.

Poor maintenance

Failure to maintain correctly may lead to premature faults, but in some circumstances this may not be the responsibility of the building owner. It could be the consequence of tenants not carrying out some of the obligations of a lease. It could also be due to the designers of the building or component manufacturers not properly (or at all) advising the owner of particular maintenance requirements.

Damage as a cause for action

A fault in a building is not in itself sufficient to provide a cause of action against anyone involved in the design and build process. For a building owner (or, indeed, a designer or contractor) to obtain redress, there must be some positive suffering. Usually this will be some form of financial loss resulting from damage.

For example, if a piece of masonry falls from a building and injures persons, property, or both, the injured party will almost certainly seek compensation from the building owner, who would normally expect his insurance to cover a claim of this sort. However, if it can be demonstrated that the fall of masonry was the result of some form of negligence there may well be attempts by the insurer to recover any cost from those alleged to have been responsible.

The actions needed once the symptoms of failure manifest themselves can, depending on the nature of the problem and the use of the building, result in enormous problems and expense to the building owner far in excess of the actual cost of remedying the faults. The expense might include not only the sums incurred directly in consequence of a claim from the injured party but also the costs of investigating the cause of the fall, and the costs of putting the building right as well as legal costs.

A fall of masonry injuring a person or persons may well turn out to be the first manifestation of a problem likely to occur elsewhere in the building unless remedial measures are taken. Sometimes emergency measures are essential, or other costly work needed to make the building safe until permanent remedial work is done. Often the latter can only be undertaken on completion of investigation work which may well be protracted, extensive and expensive.

Responsibilities

Before considering types of failure in detail, a reminder that responsibility for failure may rest with a designer, a contractor or sub-contractor, materials supplier or owner or tenant. It may also, and in the majority of cases will, lie with a combination of two or more of the above.

Physical Failure

In the area of physical failure the most prevalent symptom of trouble is water where it should not be. Other common symptoms are obvious cracking of finishes or structural elements, the detachment of pieces of cladding from the face of a building, and the corrosion or decay of materials, including rot in timber. As far as water is concerned, this includes precipitation from outside the building, condensation, ground water, water used in the construction of the building and water from leaking services. However, there are very few failures that do not have more than one cause.

Design

When design failures occur, it is sometimes argued in mitigation that it is easy with hindsight to know what should have been done. The point is, of course, that professional designers are paid to have, among other attributes, foresight. In addition to training and the experience of other designers of the same discipline, there is a wealth of information in the form of text books, British Standards (both specifications and codes of practice) and the publications of the Building Research Establishment as well as specialist trade organisations and many other authoritative bodies.

In practice, apart from the previously mentioned danger of not considering details at the stage of conceptual design, there are many reasons why things go wrong. Pressure to get a scheme off the drawing board and on to the ground is one. In a competitive world it is difficult to refuse a commission because the client insists on too short a timetable. It is also difficult at times to insist on a fee scale that will enable the proper manpower resources (both in numbers and experience) to be devoted to a project, while many clients are insistent on getting the maximum space for the minimum of outlay, regardless of the consequences. Such constraints lead to lack of proper thought and research and lack of checking.

It is also true that there is a temptation (not always resisted) for designers to design for what is seen but unnecessary, at the expense of what is necessary but unseen. In addition, many designers are unaware of the basic characteristics and behaviour patterns of many of the materials they specify, and furthermore will produce details without a full awareness of what they should be trying to achieve.

Workmanship

Dealing, briefly at this stage, with workmanship problems, it should be borne in mind that there is more to workmanship than simply the placing of brick on brick or glass in window frame. It is important to remember that setting out a building, in other words marking the precise location of elements and components of the building on the ground, as well as the height of them, is just as much a part of the building process. A site engineer who makes an error in setting out is producing bad workmanship that may have a very large adverse influence on the performance of the finished building. Another example of bad workmanship is the use by a tradesman of materials other than those which have been specified. Suppose a plumber is joining two lengths of plastic pipe by solvent welding. It would be not unknown for such an operation to be carried out using the wrong solvent for the particular plastic and for the joint to fail as a result.

Materials

As far as genuine material failures are concerned, a simple example was mentioned at the beginning of this chapter relating to substandard structural steel. Other possibilities might be underburnt bricks or timber with an optimistically given certificate of preservative treatment.

As a completely different example, consider a factory producing aluminium extrusions for, perhaps, curtain walling. Failure to check the wear on the dies used to produce the extrusions could result in problems with the finished wall. Such a situation could be regarded as a material failure, although with bad workmanship or bad management as the root cause. If the factory had supplied the extrusions to a fabricator who was either supplying to or was himself a subcontractor the contractual and legal implications of such a situation could be of some complexity! It must be said that all the examples mentioned above are rare. Unhappily, less rare are failures stemming partly or wholly from procedural errors.

Procedures

There are a number of ways in which procedural errors may result in either physical damage or financial loss or both. Dealing with an architect designed building where the architect is providing a full service to his client, the following should be borne in mind. The RIBA Conditions of Engagement state quite specifically that an architect has no authority to sanction the expenditure of money additional to that provided in the employer/contractor contract without the approval of the employer. There is an exception. The architect may incur additional expenditure on an emergency basis if safety would otherwise be at risk. Such occasions are rare. However, consider the following.

Omission of essential work

There are circumstances where essential work is omitted from the contract through oversight and the employer will accept that it is essential and would have been in the contract sum if thought about at the correct time. On the other hand, if the extra is substantial, the employer may well say: 'Had I known that the cost would have been X + Y instead of just X, I would not have proceeded with the project, and can, in any event, not afford the extra.' (It should be noted that the 'contingency' sum normally included in a contract is intended to cover the almost inevitable *minor* alterations and additions which no-one could be expected to foresee.)

Other causes of extra costs and damage

Other problems involving substantial extra costs which may or may not also result in physical damage may occur as the result of a number of factors, separately or severally. For example, late changes in detailing may result in claims by the contractor for legitimate extensions (and therefore legitimate claims for extras) of the contract period. They may also cause out-of-sequence working resulting in botched work which later fails, and there are many things which need to be considered in attempting to assess cause of and responsibility for damage. An ambiguous brief to a clerk of works may also result in the failure of either architect or clerk of works to pick up faulty workmanship in connection with components manufactured off-site. The architect may believe the clerk of works is inspecting components during manufacture in a factory, while a clerk of works would not expect to do so without having been specifically instructed. The quality of the architect's specification and details, the quality of workmanship, the correct or otherwise administrative procedures of the main and any sub-contractor, as well as the architect and clerk of works, could be singly or collectively responsible for subsequent damage.

Damage in detail

As already outlined, the basic legal premise controlling the possibility of a claim for redress is that someone has suffered damage. Also, that such damage may have far-reaching financial consequences. Before considering in more detail the consequences of damage, it should be said that physical failure may occur without causing damage in the legal sense of the word. The consequences of failure may be anything from a minor irritation which can be corrected with little cost, effort or disruption, to the need for major remedial work with many consequential costs. There have been instances where the cost of remedial work has, to use Parliamentary language of the day, approached (or even exceeded!) in real terms the original cost of the building.

The overall cost to the building owner may include any or all of the following:

(1) Temporary rehousing of the activity normally carried on within a building.
(2) Fees and builder's costs for investigation of the cause of failure.
(3) Emergency repairs.
(4) Loss of rents, profits or other income.
(5) Possibly interest on essential loans or loss of interest on capital used in connection with the litigation.
(6) Professional fees for the design and inspection of remedial work.
(7) Cost of remedial work.
(8) Legal fees.
(9) Additional payments to contractors or others due to delay or the need for out-of-sequence working or other above-normal-cost working.
(10) Depreciation in value of the property in question.

Apart from quantifiable sums there may well be considerable stress placed upon the building owner and/or employees, tenants, and others who may be involved with the owner in some related way, and such stress may well in itself have far-reaching unfortunate consequences.

It should, however, be noted that at the present time some types of consequential loss can, as a result of recent case law, be extremely difficult to try and recover, and success in an action be unlikely.

The story so far has rather assumed that it is the building owner who is the victim and the designers or builders who are the villains. This, of course, is not always the case. However, in accepting that this book is dealing with building failure rather than disputes arising from other causes, it has to be said that the most common sufferer is the owner, although tenants may be the victims of failure caused by lack of maintenance on parts of the building for which the owner may be responsible. Conversely, the owner may suffer in consequence not of design or workmanship faults, but by the failure of tenants to carry out properly the maintenance requirements of their lease.

Before closing this scene-setting chapter, two points must be made. First, because of the enormous varieties of building types, causes of failure, degrees of damage and other factors, it is virtually impossible to predict the likely costs of the consequences of failure of a building or part of a building except in the most general terms. Second, while the law in Britain follows basic principles which are slow to change, these are modified in detail by precedents established by case law. Cases are heard, judgments given, appeals against those judgments upheld or dismissed, and then possibly reversed by the House of Lords. Even without the changes that occur, the law is a highly complex subject. In Scotland and Northern Ireland there are considerable differences in both law and procedure from those in England and Wales. In consequence, legal matters will only be discussed in the following pages where essential, and in relation to routine and fact rather than opinion. Furthermore, such discussion will relate to the law as it stands only in England and Wales.

Chapter 2

Initial action following symptoms of failure

A symptom appears

When signs of something untoward appear in a building the owner or tenant may sometimes have no idea of the cause. He may know, and on occasion think he knows but be wrong. Which of the three is the case may well determine the course of action that follows. Also, they are likely to be affected by the nature and extent of any visible signs of trouble as well as the attitude of the occupier. Chapter 1 referred to major symptoms as water where it should not be, things falling off walls, cracks of varying degrees of severity and corrosion or decay.

A common scenario

Taking one possibility and following it partly down the line, consider the situation of an office building rented floor by floor on repairing leases to a number of tenants. Assume that the tenant of the top floor finds water appearing in one place on the underside of a suspended ceiling, and in sufficient quantity to fill a bucket within two or three days. The water could perhaps have started to appear shortly after heavy rain, or perhaps several days after. In the first situation the reaction of management would probably be that the water was the result of some form of roof leak, in the second either rain or a leaking pipe might be suspected. If the building is the subject of service charges there might be, depending on the ownership, a managing agent. Again, depending on the size of the building there might be no maintenance staff at all, or anything from one 'handy man' to a sizeable organisation including various tradesmen.

Initial action

Probably the first action after the bucket would be a call to the maintenance team or managing agents. Someone would remove a few ceiling tiles to look for pipes in the vicinity, and, if that did not provide an answer, go on to the roof.

From then on, there are many possibilities. Blocked outlets cleared, black sticky stuff brushed or poured anywhere suspicious in the area above the bucket, or, just as likely, a report back to the person in charge that there is nothing to be seen to give a clue to the cause of the trouble. In the latter event, and if the building is fairly new, the architect or contractor who built the project may be asked to inspect and give an opinion. The probable reaction (apart from a few unrepeatable words) will be to ask the roofing sub-contractor to have a look. After a visit by a foreman from the nearest site, there are a number of possibilities, of which the most likely are the following:

(1) The foreman finds the trouble and cures it with little effort.
(2) The foreman finds trouble in a tank room and the contractor gets his plumber to come and fix it.
(3) There are now two buckets on the top floor and the managing director has had to move his desk.

In the case of (1) or (2) the contractor will probably decide, if the building is very new, to write off the expense as a goodwill gesture. In other circumstances he will have obtained an order and be paid from service charges, maintenance fund or special levy. It has been known for a tenant on the ground floor to display little interest in a roof leak, and not be too pleased at having to contribute to the cost of repair, despite being obliged to do so. Conversely, with problems of rising damp, there have at times been similar reactions from top-floor tenants. There must be many landlords and managing agents to whom such situations will be known, relating to any multi-tenanted building. If a cure is not found, and the problems and disruption are anything more than minimal, there will be an angry tenant, or, if the building is owner occupied, a worried management.

Other examples

The manifestations of trouble might be external tiling falling from an upper floor to the street below, or perhaps in evidence on a Monday morning a bad crack of which there was no sign on Friday evening. Assume, however, that whatever the symptoms, the occupants/ owners/agents/local or original contractor know neither the cause nor how far and how quickly it is likely to spread or the consequences if it does. Also assume that someone has complained to the owner who has decided that the problem is going to involve some expense to put right. Provided the building has been recently completed and the owner is on good terms with his designer, the former will probably approach the latter with a request to investigate and advise. There are a number of possible responses. From the author's own experience, any of the following would not be unusual:

(1) The designer finds the source of the problem by physical inspection, and diagnoses, correctly, a latent defect in work carried out by the contractor or a sub-contractor. The fault is put right at the contractor's expense, after perhaps a degree of discussion and negotiation.

(2) Same beginning, but the contractor refuses to accept responsibility, and claims that the problem was caused by a design defect.

(3) The designer correctly diagnoses that the problem has been caused by poor detailing. He admits the fault is his responsibility, and pays for the error to be corrected. This very seldom happens unless the money involved to correct is very little, and in fact much less than the excess on the designer's professional indemnity insurance.

(4) The designer fails to diagnose that the fault is due to poor detailing, and tells the owner that it is:

(a) due to poor workmanship
(b) due to the building 'settling down'
(c) 'must be faulty materials'
(d) poor maintenance or something similar, perhaps misuse by tenants or employees.

   The reasons given are as likely as anything else to be the product of wishful thinking.

(5) The designer does not know the source of the problem and suggests that the owner employs a specialist consultant, or, depending on size of practice, reputation and sense of responsibility, may employ one himself.

   There are on record cases of designers wrongly convincing their clients for long periods, even years, that failures are not due to their design faults, in the great majority of cases (but regrettably not always) because that has been their genuine view.

The original designer

There are reasons why a building owner might not seek advice from the original designer. Relationships may have been soured during the course of the work, the designer may not have been providing a full design, inspect and certify service, the building may have been 'sold on' by the original developer, or the designer may no longer be in practice.

Where, however, the original designer is available and the owner sees no reason not to refer to him, he should still seek legal advice before an approach is made. In some circumstances it may be undesirable to alert the designer to the fact that there are substantial problems which could be his responsibility. At other times obtaining an opinion which may subsequently reduce the designer's room for manoeuvre could be a positive advantage.

Finding technical help

The owner, faced with a problem, and unable or unwilling to turn to the original designer, may, depending on the nature and severity of the symptoms of failure, either deal with the matter on an emergency basis or be able to consider first in detail the best course of action to take. His first approach is likely to be to a local professional practitioner in the building industry, such as an architect or building surveyor, or perhaps to a local contractor in whom he has confidence. Any such persons may have been recommended or may be known to the owner. Sometimes, even with a small building, if the problems are severe, the owner may ask for advice from one of the representative bodies such as the Association of Consulting Engineers (ACE), Institution of Civil Engineers (ICE), Institution of Structural Engineers (IStructE), Royal Institute of British Architects (RIBA), Royal Institution of Chartered Surveyors (RICS), Faculty of Architects and Surveyors (FAS) or the Chartered Institute of Building (CIOB). Another first contact might well be the Building Research Establishment (BRE) Advisory Service, or possibly one of the trade federations related to the apparent problem. Note, 'apparent'!

With regard to approaching another professional for help, it is not considered ethical by most professions for a member to become involved with the work of another practitioner still in any way connected with the project. It is normal to take active steps to ensure that this is not the case. An exception with some professions is if there is a genuine prospect of litigation by the building owner against the original designer.

The amateur 'expert'

One of the problems in dealing with building failure has an analogy with medicine. Most people at some time or another have come across the situation in which a person suffering some symptom either believes that he knows himself what the trouble is, or has a well meaning friend, relative or neighbour who knows exactly what the trouble is, and what is more, how to cure it. Even worse, and not unknown, doctors themselves have been known to be wrong and failed to be alerted to the need to obtain specialist advice until considerable damage has been done, or suffering endured, by the patient.

Fortunately as far as the human race is concerned, the latter state of affairs occurs far less frequently than it does with buildings. It is true that many building problems are simple to diagnose, and there is no need for specialists to be involved. Perhaps the most important single fact to remember in connection with building failure (and a great many other things as well) is for anyone seeking the answer to a problem to be aware of their own limitations, and be prepared to admit them to others.

If it were possible for this policy to be followed without exception not only would a great deal of time and expense be avoided in the correction of failure, but far less failure would occur in the first place. Unfortunately, apart from carelessness, arrogance and incompetence there are always people who *know* that parallel is spelt parralel and that two and two make five. In consequence, while there are times when the general practitioner (be he architect, engineer, surveyor or whatever) will suggest the employment of a specialist, there are others when diagnosis and subsequent remedial work are undertaken without a proper understanding of the problems and are consequently ineffective. Even worse, the remedial works may be undertaken with no diagnosis even attempted.

In this context it is necessary to appreciate what is meant by specialist. Unlike the medical profession, this usually means not someone dealing with one particular branch of building, but someone who has specialist experience of or aptitudes for the investigation of failure. It may also be necessary, after an initial appraisal, to call in specialists in a particular part of the building process, be it structure, asphalt, mortar, paint formulation or some other aspect of the construction.

First action in detail

To look in more detail at the actions that should be taken in the event of symptoms of failure, broadly speaking, buildings may be subdivided into small and large, problems into moderate and severe (it is assumed that minor problems can be dealt with without recourse to experts, although if they escalate, they may need to be re-assessed), and their type perhaps ranged from simple to complex.

Small buildings

Considering first small buildings, these might be single houses, houses converted into two or more self contained dwellings, other smallish residential buildings, including perhaps private nursing homes, small office buildings and workshops. There are many other examples of a similar nature, probably owned by individuals or small businesses.

Initial action by the owner, depending on the nature and severity of the symptoms of failure, would reasonably be to approach the original designer or contractor if known, otherwise, a local architect, engineer or building surveyor depending on the type of problem.

Whomsoever it is decided to approach it is important that, even with what appears to be the simplest of problems, the person appointed is a suitably experienced professional. Before appointment they should be asked to state their experience of such problems and demonstrate also that they possess suitable professional indemnity insurance covering any successful claim for negligence. This assuming of course that their advice is other than to recommend approaching a specialist if one is known to them or the appropriate professional body if not.

If the professional is himself happy to diagnose the defect and subsequently design remedial work he must be asked at the outset if he is prepared to act as an expert in the event of the failure proving to be the result of actions of someone other than the owner, and in consequence, litigation or arbitration is contemplated.

He should also be asked if he is fully aware of the duties and responsibilities of an expert if litigation should, in fact, subsequently be embarked upon, and prepared to act as such.

It is essential to use as an expert someone who will have the necessary expertise to be able to diagnose and correct the faults. If a suitable local practitioner can be found, well and good.

Still on the subject of small buildings, but considering a more complex problem with the causes not immediately obvious to a general practitioner, it may be possible to save considerable sums by using a specialist to carry out an investigation and diagnosis and then act as an adviser with a local practice doing the detailed design and follow-up work to completion. This, however, should be a last resort and undertaken only after very careful discussion and agreement regarding the division of responsibility. This is essential if problems resulting from misunderstandings over duties and design intentions are to be avoided.

If an investigation leads to the conclusion that the failure could have been caused by anyone other than the building owner, legal advice is essential at a very early stage, if only to establish the likelihood of being able to recover the costs of the remedial work.

Large buildings

With larger buildings it is advisable unless, again, the problem is a minor one and the cause obvious, to seek specialist advice from the outset. It was suggested earlier that the initial source of information to identify suitable experts might be one (or more) of the professional institutions, trade associations or the Building Research Establishment who operate BRAS (Building Research Advisory Service). This organisation not only employs people with expertise in a number of aspects of building, but also has the backing of technical services, including laboratory and testing facilities, and will, for a fee, carry out investigations and give some advice. As a Government department, however, they have in the past been found, understandably enough, unwilling to act as experts for one party to a dispute. They are also usually unwilling to give other than general advice regarding remedial work.

The majority of practitioners specialising in the investigation of building failures will not only act as experts as far as diagnosis is concerned, but are also prepared to give evidence in court or to an arbitrator, and will also accept responsibility for the design and implementation of remedial work.

Responsibility for failure

More often than not it happens that, whoever carries out the initial investigation, the building owner is advised that people other than himself have been at least partly (if not wholly) responsible for the problems. Be it designer, contractor, tenant or other, the owner, not surprisingly, decides he wants to try and recover monies already spent as a result of the failure, and cover the cost of any relevant future expenditure. The chances are he will consult the solicitor he normally uses to assist him with other business.

It is vital for the building owner that all his advisers know, not necessarily what they *are* doing, but what they *should be* doing. Unless the owner is already certain that his legal advisers understand building litigation, he should ask the question, and be sure of an unequivocal answer that gives him the assurance he must have. (As unequivocal as the answer he should have received from his potential technical adviser.) He must be sure that all his advisory team have the specialised knowledge needed, or, at the very least, the resources and the will to acquire it, including seeking help and advice from any other specialist source that may be necessary. This is another situation where it is essential for all concerned to recognise when they need further advice.

Choosing legal advisers

The Law Society publishes regional directories and there is also a list which gives the special expertise of local firms in the first case, as well as large practices with specialist departments. This includes a category related to building law. Incidentally, most of the professional institutions issue directories of members and/or practices, some of which give details of firms claiming to have specialised knowledge of the investigation of building problems and the design of remedial work as well as experience as expert witnesses. Such directories, including those published by the Law Society, may be found in public libraries and Citizens Advice Bureaux.

One way of obtaining expert legal and/or technical help is by recommendation from someone who has experienced, and dealt with, building failure problems in the past. However, reference to appropriate professional bodies is perhaps the safest if a reliable recommendation is not possible from other sources.

Once a legal adviser has been found he will usually know of suitable technical experts, and vice versa. It will be appreciated that if, after consulting both technical consultants and solicitors, it is decided to take legal action, if the situation is anything but the simplest, it will be necessary to obtain opinion from counsel, who as a barrister will also plead on behalf of his client should the case ultimately reach the courts. As with other professionals, there are barristers, both QCs and their less senior colleagues, who specialise in building law. Their nomination will normally be as the result of a recommendation from the solicitor. At the time of writing this dual system is a matter of fact. Time will tell whether the widely publicised proposals to change it come into effect.

The need for and timing of the expert's appointment

It often occurs that investigation and diagnosis, and sometimes remedial work as well, is carried out by someone who is either not prepared to act as an expert witness in a law suit, or possibly is not considered by the lawyers as having sufficient experience to do so. Sometimes an expert is not sought or appointed until litigation has become a certainty with the belief that this will save money on experts' fees. In consequence, it frequently happens (and all too frequently) that by the time the expert is appointed valuable evidence has been lost or destroyed, and neither the owner's nor indeed anyone else's experts can form an accurate picture of the symptoms, the failures and their causes. This can have disastrous consequences as far as the outcome of litigation is concerned.

It is readily understandable that building owners, faced with what may appear to be simple, relatively trivial problems, are reluctant to engage experts in the mistaken belief that they are expensive in terms of their fees and the recommendations they make. It is equally understandable that owners are reluctant to take similar action with regard to obtaining legal advice. It is essential not only to obtain first-class technical and legal advice but also to obtain it as early as possible. The reasons will become clear in the following chapters, but the principal ones are summarised now. First, it is emphasised that there are 'High Street' practitioners, both technical and legal, capable of dealing with building failure in respectively, both construction and legal terms. However, as already noted, they *must* be aware of the limitations of their skill, experience and resources in order that they keep their activities, even if stretched (not necessarily a bad thing), within their capabilities. Even before that, though, is the equally important need for the owner himself to appreciate that the building has problems and that they are beyond *his* skill and experience to sort out.

Like the young surgeon performing his first operation, the technical expert has to start somewhere, but is normally someone with considerable experience of his own discipline. He must possess very definite qualities to perform the work efficiently. The solicitor needs to have or acquire his specialised knowledge of law and practice relating to this field as well as the remaining attributes which are more or less taken for granted. The same goes for the barrister.

More than one cause of failure

Another reason for the early appointment of a suitably qualified technical expert is that there are often several causes for the problems encountered. Furthermore, they are very often not those that appear to be the obvious ones. In addition, an experienced expert, while not a lawyer, will be able to give a reasonably accurate assessment of the degree of responsibility of each of the parties involved (including tenants or even the owner himself) and will be able to say whether a claim is likely to succeed as far as the *technical* issues are concerned and the duties which members of his profession should have undertaken, and how they should have been undertaken.

On the other hand, lawyers with appropriate experience will be able to give an indication (at least in terms of the law as it stands at the time they are consulted) of the likely legal outcome of a claim or result of an arbitration.

With both technical and legal experts, such opinions will be based, of course, on the information available at the time they give the opinions.

Discussions between client, lawyer(s) and expert(s) at a comparatively early stage will enable an objective decision to be made on the likely merits, including financial, of proceeding with a claim. In fact, calling in the right people at the right time and allowing them to do their job in the early stages will, in the great majority of cases, result in very considerable financial savings, as one of the most important functions of both technical and legal experts is to tell their clients not only when they have a good case, but when they have not. That holds good for both the potential plaintiff and the potential defendant.

It must be emphasised that the litigation process from initiation to conclusion can take several years, and the law relevant to a particular action can change, as a result of judgments, when much time, effort and money have already been expended. Arbitration can sometimes be a quicker process, but has its cons as well as its pros. Nevertheless, the majority of cases are decided on the basis of the law as it stood at the time the action was first embarked upon, as changes brought about by case law or new legislation are not that frequent.

Issue and service of writs and limit of liability

There is another extremely important matter relating to the timing of an action. There are limitations on the periods between the occurrence of damage and the initiation of action. These are discussed later, and at this stage it is enough to appreciate one major reason for the early issue of a writ by the owner on those thought to be responsible for the building defects. It is possible to *issue* a writ recorded by the courts which is not immediately *served* on the relevant party. The actual issue has the effect of freezing the time scale within which action has to be taken for a maximum of 12 months during which the writ must be served or lapse and become ineffective unless renewed. This is useful in situations where the cause of action might otherwise become 'time barred'. The limitation between occurrence of damage and issue of a writ is a complicated subject. The period varies from 6 up to a maximum under present law of 15 years. This limitation is discussed in Chapter 7 under the heading 'Legal grounds for action'. It will be seen therefore that if damage has occurred five and a half years before an investigation is carried out, the issue of a 'clock-stopping' writ could in some circumstances buy vital time. The timings mentioned above are a massive oversimplification of a complicated legal situation which has not been resolved by either the Latent Damage Act of 1986 or subsequent cases which have established precedents creating 'case law'.

Few cases are clear-cut, and sometimes the decision to proceed or not with litigation is a difficult one. Sometimes commercial settlement decisions are made which, while not necessarily resulting in a fair outcome, are the sensible financial answer. These may take into account the financial situations of any or all of the parties as far as they are known or can be established.

Whatever the situation, and whether litigation or arbitration is being considered, one thing still holds good: the earlier someone with a problem obtains proper advice, the less likely they are to waste their time, energy and money, and the more likely they are to recover the bulk of their costs.

Chapter 3

The role and briefing of technical experts

Why use technical experts?

Technical experts are needed in connection with many different types of legal process, and often their investigations and findings will dispense with a need for legal or arbitration action *in toto*. For example, an investigation may reveal that a failure is the consequence of a natural phenomenon for which nobody can be held responsible, or, alternatively, the evidence of responsibility is so clear-cut that argument is fruitless. Experts may be found in the sphere of medicine, shipping, aeronautics, other forms of transport, many branches of engineering, computer science and a host of other fields.

The principal reason for their initial employment is to establish with the aid of the specialised knowledge, experience and techniques outlined earlier, the cause(s) of a failure. From that point on, experts also have the knowledge of contractual obligations to give opinions on the responsibilities for the failures where such responsibilities are more than purely technical issues. They have the capability and experience to design remedial work and see it satisfactorily through to completion.

If litigation or arbitration are chosen as the way forward the expert will present an objective report, whichever party he is acting for. Experts for all parties will meet and agree as much as can be agreed to reduce the issues in contention and in so doing reduce court or arbitration time and therefore costs.

Who are technical experts?

In connection with building failures the principal experts are usually architects, structural and services engineers, building surveyors and sometimes quantity surveyors. They will frequently need to call on the services of specialists such as chemists, physicists, timber experts, etc. Normally, an expert will be a person practising his particular occupation and possessing considerable general experience of that occupation and related fields.

Essential attributes of the expert

It is most important that the expert should possess, apart from experience, the following qualities:

(1) An enquiring, analytical and practical mind.
(2) Thoroughness and complete objectivity.
(3) An eye and ear for the essentials of a situation.
(4) The ability to treat every problem strictly on its merits. It is most important for an expert to avoid the trap of thinking 'This is the same as so-and-so'.
(5) A knowledge of information access and retrieval.
(6) The ability to use clear, concise, jargon free language, both orally and in writing, supported by clear, simple drawings and photographs.
(7) A good memory and the ability to think clearly and quickly under pressure, including that created by opposing counsel.
(8) The acceptance at times of tight deadlines related to the production of information.

Tact and diplomacy are also desirable qualities.

The 'GP' expert

The above apply to experts of all disciplines, but in the normal course of events most of the work will be done, as far as buildings are concerned, by architects, structural or services engineers or surveyors. Each should have a basic knowledge of the work of the others and a basic (for the architect) knowledge of physics, mechanics and the behaviour of materials in use. In the case of the architect a sound knowledge of building construction and site methods are absolute 'musts'. These are of course all attributes that should be possessed by any normally competent architect. The cynical reader may recall however, that some 60 per cent of building failures are the direct consequence of design faults.

The term 'knowledge of information access and retrieval' in point (5) above is a most important part of the expert's work which occurs at various times during the investigation, report writing, and preparation for trial stages and frequently, if not invariably, during the actual hearing of a case. It is stressed too that 'objectivity' covers, apart from anything else, a need for an awareness that it is easy to make mistakes, and easier still to be wise after the event. In addition, while compassion and sympathy may be extended to the perpetrator of a fault, that fault has in the majority of cases caused damage to someone else who is also deserving of sympathy. Perhaps of all the qualities needed by the expert, those of detachment, objectivity and impartiality are paramount. To those who say 'I am paying an expert to look after my interests', it is as well to point out that it is in the interests of any party to a dispute to know and act on the weaknesses just as much as the strengths. There is no sense in proceeding with a case without a reasonable chance of winning it. There may of course be situations where a bad technical case may succeed on legal grounds, and vice versa, but whatever is done must be based on a true assessment of the entire situation.

What does the expert do?

Essentially, the function of a technical expert acting for a building owner with a defect or failure problem is to investigate and establish its cause or causes. Having done so, the expert may then be required to advise on the nature, cost, timing and method of execution of remedial work. There will probably also be a request to advise on the question of responsibility for the failure, perhaps combined with advice on the merits or otherwise of seeking recompense for damage suffered. Certainly this is so far as technical issues are concerned, although there is frequently an overlap between technical and legal aspects. Should legal or arbitration action be proceeded with, the expert will invariably then be called upon to act as a witness during the hearing. As an expert, his evidence will be facts based upon a great deal of information which he will acquire and analyse during the progression of his work and opinions based upon those facts. Apart from giving evidence, he has an additional advisory role, dealt with in Chapter 11.

Opposing experts

The expert(s) acting for the person or persons alleged to be responsible for the problems (potential defendants) will have a similar role in some respects to the owner's expert. This will be principally to assess all the evidence they have themselves seen or otherwise accepted, presented by the plaintiff and any other defendants and/or third parties (a third party is someone brought into an action by a defendant rather than by the plaintiff as being considered partly or wholly responsible for what has transpired) to see to what extent such evidence appears sound and to offer their own opinions upon it. Sometimes there is agreement between experts on certain aspects of a failure. Very occasionally, this agreement is substantial. If all experts were completely objective all the time substantial agreement might well be quite frequent! In fact, the incidence of debate and disagreement between informed persons on a multiplicity of topics in varying walks of life suggests that complete objectivity and near identical interpretations of fact on any particular subject occur less often than might be hoped. Hence the widespread occurrence of litigation and arbitration.

Briefing by a building owner

Usually, the brief to a technical expert will start with a request either to investigate a problem or to confirm someone else's diagnosis with a view to litigation or arbitration. For reasons which have previously been briefly mentioned and which will become increasingly obvious, it is far more satisfactory if the expert is called in right at the beginning rather than later simply to 'deal with the litigation side'. This latter situation happens all too often.

First requirements

Assume that the owner calls in a suitable person when the symptoms of failure first appear. Normally, the initial request would be to give an indication of the cost in fees and expenses of investigating the symptoms, finding their cause(s) and, usually, making in addition some sort of stab at the consequences in terms of money, time scale, disruption and correction of the trouble. In the majority of cases however, after a brief initial inspection the expert would only have a very approximate idea of likely fees. It is most unusual for an expert to work on anything other than time charges, simply because it is almost impossible to assess, until at least some work has been done, how much time will still need to be spent on arriving at positive conclusions. While remedial work is often done on a percentage scale fee, the expert has no idea at the outset of his work of the reasons for the trouble, therefore, of course, no idea of the remedy and no idea of either the cost of the work or the associated fees.

Fee control

Once an initial inspection has been done it may be possible to give some indication of fees to the stage of indicating causes and possible remedies. There are dangers consequent upon the potential client insisting on a fixed fee either from one suitable practice or by shopping around for quotes from several. First, the expert may pitch his fee bid too high to ensure not making a loss. Alternatively, he may pitch it too low, and then be tempted to cut corners to try and avoid a loss, which would be disastrous. Cutting corners as a result of accepting a cut fee is sometimes a cause of design failure, and cutting corners in the case of the investigation of failure is a certain recipe for disaster for both client and expert. An 'expert' who is prepared to work for a cut fee is, except in the most exceptional circumstances, not worth employing. (The exceptions may include such situations as work for worthwhile charities or individuals who just fail to qualify for legal aid.) It is absolutely vital that the investigator is allowed to do his job properly. If an accredited professional is employed, his fee will be earned by the quality and objectivity of his advice. This might be, in conjunction with legal opinion, that there is real hope of recouping the bulk of the costs consequent upon the failure, or, conversely, that to try and recoup the cost would be a waste of money.

A method of fee control

There is a way that is frequently used to avoid the expert being given a blank cheque, and the sequence of events is this:

(1) Agree with the expert hourly rates for principals and such of his staff as are likely to assist him.

(2) Agree an initial inspection to give the expert an opportunity to assess the likely magnitude of the problem and time in hours that will be needed for those involved. This estimate of time (and therefore fee) is bound to be very approximate but will at least give the client some indication of how many noughts will be needed to follow the figures on the cheques.

(3) Arrange for frequent updates of hours spent, having agreed intermediate fee 'ceiling' figures, of which the client has to be advised as they are approached.

The owner has one assurance with regard to expert fees. The

majority of good experts experienced in the investigation of building failure are usually too busy to spend more time on an investigation than is strictly necessary. The great majority of recognised experts however, can usually be relied on not to cut corners because they are busy (although this is not a guarantee!). Most investigators will tell a potential client whether or not they are in a position to cope with a commission, although they may not be in a position to do more than indicate the capacity for an initial inspection until that inspection has been made and the likely future work load assessed. This incidentally is another reason for the owner to appoint his expert as early as possible.

Timing, early actions and other parties

It was stated earlier that the initial brief to the expert might well be simply to establish the cause of failure. It may be, even before that, to make a preliminary inspection to enable some estimate to be made of the likely time needed for an investigation and report on the cause of failure in the first instance and the hours and fees to be approximately estimated if possible. It must be appreciated that, for example, 30 hours' work may well be spread over several weeks. This could be for a number of reasons. For one, other work already committed which has to be fitted in. Also, samples may have to be sent for analysis, photographs processed (although the initial processing is quick, several copies may need to be obtained, mounted, captioned, etc.). As well as the actual writing, processing and assembly of the report itself.

Intimation to original designer

When architects and engineers are asked to do work in connection with a building designed by someone else it is a normal requirement of most professional bodies that members should clear with the original designer that the appointment of the latter has been properly terminated. There is no need for this to be done if there is any prospect of litigation arising out of an investigation. As already suggested, there may be advantages in asking the designer for an opinion at an early date, but legal advice should be sought before an approach is made.

Presence of others at initial inspection

At the time that the first inspection is made, the expert will do his best to assess the likelihood of other people being involved, such as the original architects, engineers and/or contractors. This is highly desirable, as a detailed investigation will, in the majority of cases, need to be carried out when those who may be held liable have been advised and given an opportunity to be present or represented. If in fact it appears likely that a claim will be made by the building owner the expert will also need to discuss further action with both the owner and the latter's legal adviser. Either his usual solicitor if suitably experienced in the building field, or, if this is not possible, a firm recommended as being suitable to handle a building dispute.

Initial 'opening up' on site

After the initial inspection and assuming the owner has agreed the need for a full investigation, this will be carried out either by his own expert only, or with other parties experts' attending as observers. Certainly at this stage, and possibly even at the preliminary investigation, there will be a need for some opening up of the building fabric. This opening up is usually done by an independent contractor working under the direction of the expert. The costs of this opening up and subsequent 'making good' will need to be allowed for in addition to the expert's fees. Such work can seldom be done except on a 'day work' basis, i.e. the cost of materials and labour plus a percentage for overheads and profit. Depending on the size of the problem and the nature of the work this could be anything from a few pounds upwards. This stage will normally be followed by an analysis of the evidence to establish causes.

Temporary remedial work

If the owner wishes, the expert will, after preparing his diagnosis of the trouble, prepare a preliminary scheme for any essential remedial work. (Note *essential*. While a building owner may take the opportunity of doing remedial work to carry out other desirable work, the cost of non-essential work cannot reasonably be added to a claim.) Such proposals, together with an indication of approximate cost, would be necessary in order to reach a decision regarding further action, quite apart from the cost of any short term emergency repairs needed to protect the fabric of the building and its occupants and contents.

Detailed investigation by owner's expert alone

It may be that initially the owner's solicitors do not wish to indicate that a claim is likely, and may ask that detailed investigation be carried out without other parties being advised. If this is the case it is highly desirable that enough areas remain that can subsequently be investigated in the presence of other experts. Only if full remedial work is necessary on an emergency basis before the allegedly responsible other parties are advised is it sensible for it to be undertaken before inspection by the other parties' experts. Nevertheless, there is no legal obligation on the part of a potential claimant to invite potential opponents to be represented at site investigations.

Records of investigations

In such circumstances, the most meticulous records must be made of the investigations carried out solely by the owner's expert. In fact, comprehensive and accurate records are absolutely essential whatever the circumstances. These would normally take the form of photographs, sketches, notes and records of moisture readings and similar relevant matters depending on the type of symptom. Video recordings are sometimes made but their use during an investigation can be a positive disadvantage, with the 'cameraman' getting in the way of both tradesmen opening up and investigators trying to take instrument readings and other essential measurements. If used, a considerable amount of experience and preplanning of what is to be done and how are essential.

Informing owner and other parties

All the matters described above should be drawn to the attention of the building owner by his legal and technical advisers, as appropriate, before the expert's brief is finalised as they may affect the owner's instructions to his expert. Other issues to discuss are the importance of ensuring if possible that the other parties are advised as early as practicable of impending investigations in order that if early and extensive remedial works are essential there can be no suggestion that evidence has been destroyed. However good the records of an investigation, the knowledge that other parties have seen the same symptoms and the state of the building that has caused them is absolutely invaluable. This applies whether the failure has been caused by poor design, poor workmanship, poor maintenance or a combination of two or more. (Usually problems arising principally from poor maintenance will be apparent during a preliminary examination, before anyone but the owner's expert is involved in investigation.) Apart from anything else, it will reduce the likelihood of one party attempting to play off deficiencies against another, as all will have seen the true state of affairs, even if they draw different conclusions from what they have seen.

Further briefing

In most cases, unless the problems are extremely simple to deal with, once the causes of the problems have been established the building owner, probably by now established as the plaintiff, will require additional services from his expert.

The design of permanent remedial work

Apart from continuing to act during the preparations for trial and during the trial itself, the expert will probably be required to advise on the detail of remedial work and its implementation. There are instances where remedial works are not designed by the expert, usually for one or more of the following reasons:

(1) The plaintiff or his lawyers feel, or think a court may feel, that the expert may recommend more remedial work than is really necessary in order to create extra work for himself.
(2) The expert may himself not wish to do the remedial work, either because he does not have the resources or because it is his policy not to do so.
(3) The expert's office is located a long way from the site and the owner feels that site inspection and contract administration might suffer in consequence as well as being more expensive.

In fact, where other considerations are equal, it is preferable for the expert to design and organise remedial work for the following reasons:

(1) Matters relative to the case uncovered during remedial work are more easily brought 'on board'.
(2) The expert will have a greater knowledge than anyone else of the pitfalls that may be found and the ways of overcoming them.

(3) If the findings and suggestions of the expert are used by another practitioner there will almost inevitably be problems of divided responsibility between the two of them for the finished work.
(4) Opposing experts will be looking closely for non essential remedial work, whoever designs it.
(5) Experts worth employing are invariably sufficiently reputable to want to avoid on ethical (and common sense) grounds any possible accusation of doing unnecessary work.

Briefing by a potential defendant or third party

Once someone has decided to try and recover financial loss due to damage, the party or parties allegedly responsible for the damage will at some stage be presented with a claim, perhaps in the form of a 'letter before action', or sometimes the first notification may be the service of the writ.

Either way, once a claim has been made the recipient will normally take legal advice on the steps to be taken, and if there is an intention to defend and the plaintiff is relying on expert advice and evidence (and sometimes even if he is not) the defendant will invariably, sooner or later, appoint an expert to assist him. In the early stages such appointment will probably be to examine the technical case of the plaintiff as set out in the claim and give an opinion on the merits of the technical issues presented. Even in the early stages it is likely that there will need to be consultation between the defendant and his legal advisers as well as the expert. The latter may have been recommended by the lawyers (solicitor or counsel if briefed at this stage) or possibly be someone already known to the defendant, either personally or by reputation. All too often defendants, and sometimes their lawyers too, attempt to go too far with the preparation of a defence before calling in an expert, and may prejudice their case due to the covering up of useful evidence. Frequently this state of affairs occurs because not only do the lawyers lack the technical knowledge needed, but so too does their client!

Further briefing by the defendant

Once the expert has made an appraisal of the information available, his advice may be to resist, to pay up, or, that he cannot make a positive recommendation without further information. The last of these may well be dependent on the further action taken by the plaintiff. If his advice is to pay up, he may well be asked to give further advice on how much to pay. This will almost certainly be in conjunction with the lawyers and possibly quantity surveyors. It is also possible of course that the defendant may ignore suggestions that he has no case and decide to fight anyway. In such circumstances the technical and legal experts will need to qualify their further role or withdraw if they think that to proceed with a defence is a waste of time and money. This type of appraisal in fact needs to be made more than once, as more information about claims and evidence becomes available, so that there are regular updates relating to the chances of a successful defence, assuming of course that the dispute does proceed.

On the assumption that it does, the further activities of the defendant's expert will be controlled to a considerable extent by the plaintiff's advisers. His future efforts will be directed towards attending site investigations, recording data and observations made during the investigations and preparing his own report based on those investigations as well as all other available information. Eventually, should the case proceed to either arbitration or the courts, he will perform a similar role to the plaintiff's expert but on behalf, of course, of his own client.

Fees

Inevitably, as in the case of the plaintiff's expert, fees will be on a time-charge basis. Once the decision to defend an action has been taken there is little that can be done to control fees as the amount of work needed will depend to a large extent on the investigations arranged by the plaintiff's expert and the amount of documentation and other evidence that has to be assessed before a report can be finalised. What in fact all parties' experts will do, wherever possible, is arrange for the gathering of factual data to be done by relatively junior staff. Obviously, however, this must be under the direct control of the expert himself, and the staff must be sufficiently qualified and experienced to ensure the validity of the findings they record. It is usually only in the case of straightforward recording of repetitious work that this can be done. Even then, briefing and supervision must be beyond reproach. Also, the expert must ensure that he personally sees a representative sample of the evidence gathered by others. The defendant's or third party's experts will not be engaged in the design of actual remedial work, but will almost certainly be required to check the designs of whoever is employed to do them. In some cases, the defendant's experts may have to prepare remedial schemes if they are convinced that more satisfactory and economic solutions can be achieved than those produced by the plaintiff's chosen consultants. In court, the role of the defendant's and third party's experts will be very similar to that of the plaintiff's expert.

Budgeting for fees

Anyone proceeding along the litigation or arbitration roads is presumably doing so because he believes he will win or can achieve a substantial and worthwhile settlement. In this regard it is important to remember that at the start of a case litigants on both sides tend, in the main, to be convinced that they have right on their side. In the majority of cases judgments do not result in the award of the total claimed, and whether judgment is for or against the plaintiff, it is unlikely that any party to a dispute will recover all his costs. Add to this the knowledge that optimism by everyone involved also tends to extend to the amount of time that will need to be spent by the parties themselves and to their technical and legal advisers. Such optimism should be discouraged, and the worst allowed for. Happier the pleasant surprise of discovering all is better than feared instead of worse than hoped for. Nevertheless, people do frequently, at the end of the day, decide it has been worthwhile!

Extent of briefs for all parties' experts

Whether acting for owner, contractor, designer or any other involved party, the expert has a very clear duty, before he is commissioned, to explain to his potential client as accurately as he possibly can the likely extent, from a pessimistic viewpoint, of his involvement. The work to be done, the likely time scales and some approximate indication of fees, other costs and his own involvement should be spelt out in considerable detail. To be avoided for the sake of all concerned is the situation where the potential litigant reaches a point of no return with no real knowledge of his commitment whether he wins or loses. A comprehensive and detailed brief is, in many cases, best produced by the legal and technical experts and ratified by the client after any necessary explanation and/or amendment. Briefs should preferably be taken one stage at a time, provided that the client is fully aware right from the beginning of the likely (there is always the unforeseen) consequences arising from each succeeding stage.

Chapter 4

The expert's investigations

Chapter 3 touched on the investigations by the owner's expert. There are occasions when the failures are of such a nature that the expert can obtain all the information he needs from an examination of documents, but these relate chiefly to loss or damage caused by contractual or procedural problems rather than physical failure. They may also be all that is needed in cases where the principal expert is of one discipline, perhaps an architect, and one of the defendants is, say, a structural engineer. In such instances a structural engineer may be required to deal purely with the duties of the engineer to his client rather than technical issues, and may be able to prepare his report entirely from a perusal of documents in his own office.

Taking as an example a similar situation, assume that an investigation has been completed by a building surveyor into a building designed by an architect. The surveyor says in his report that in his opinion some of the responsibility for the defects in the building lies with the architect. There could be a subsequent suggestion by opposing counsel that the surveyor is not qualified to deal with the duties and extent of reasonable knowledge and skill of an architect. To avoid this situation an architect expert may well be briefed to deal purely with such aspects of the case, and without having been involved with the gathering and interpretation of evidence.

Types of investigation

Investigations may be divided into the following categories:

(1) Site investigations on or in the building.
(2) Examination of relevant documents.
(3) Laboratory examination of samples removed from the building for scientific analysis.
(4) On-going site inspection and recording during the progress of remedial or other work.
(5) Relevant research.

Site investigations

Site investigations may be subdivided into a preliminary look merely to get an indication of the scope of the problem, preliminary 'opening up' to commence the analysing process, full scale opening up and detailed examination and on-going investigation/recording during remedial works. The last two will almost certainly be attended by other parties' experts as observers, and either of the first two may be. It must be realised that at any stage there may be a need for the expert to be accompanied by specialist experts of other disciplines. For example, if the expert is an architect or building surveyor, he may need advice at certain stages from a structural engineer, an industrial chemist or perhaps a foreman asphalter. The need for these specialist examinations may not always be possible to foresee until some preliminary work has been done. Frequently however, specialist advice can be limited to examination of photographs and drawings, and sometimes even a phone call is sufficient. Each case has to be treated on its own merits.

Photography

Some experts tend to use professional photographers. There are disadvantages to this. If other parties' experts are present it is not always desirable to show one's hand by spelling out to the photographer precisely what the photograph is to highlight, as trying to do this without drawing everyone else's attention to the prime subject is not easy. Anyway, it is much easier and quicker to teach a building expert how to take suitable photographs than to teach a professional photographer building construction! Also, the expert taking his own photographs saves a great deal of time, particularly if, as is likely, all the experts require photographs. Each of them briefing a different professional photographer would border on farce, particularly bearing in mind the practical difficulties of access sometimes encountered.

Video recordings

Some experts favour the use of videos as an additional record. However, they cannot completely replace the use and value of photographs which can be readily referred to and looked at in detail without the need constantly to stop and start a video recorder. Furthermore, the use of video cameras on site, especially if more than one person is each trying to take his own shots, is extremely cumbersome and in many cases really more bother than it is worth. This does not mean that in some circumstances they may not be of use, nor does it preclude the use in court of video recordings made elsewhere which may have a bearing on the subject under discussion.

Organising a site investigation

The arranging of a site investigation can of itself be quite a complicated exercise. If, for example, local authority housing is the subject of the investigation, there will be a need to notify housing managers, tenants and others, and obtain identity cards for the expert and those assisting him. That is quite apart from organising access to various parts of the building, perhaps roofs and plant rooms. Ladders, access towers or other plant may be needed. There may be a requirement for scaffolding and at all times the provisions of the Health and Safety at Work Act have to be thought about and observed.

Opening up and recording evidence

The expert will have to ensure that he has adequate assistance and the equipment necessary to record everything that may possibly be relevant. Opening up requires skilled labour, tools and, absolutely essential, the wherewithal to make good after an examination. It has been known for a less than competent investigator to go on a roof that has five leaks and leave it at the end of the working day with seven! The experienced and capable practitioner will however be alive to this possibility, and also know what to look for once an initial appraisal has been made. It is vital to ensure that each inspection is treated wholly on its merits and nothing is taken for granted. However, the correct approaches to site investigation ensure not only adequate closing up of areas investigated but also reduce on site time and disturbance to the building and its occupants to the minimum.

Not only is meticulous recording of information (including anything that could possibly be of importance) vital, but so is adequate sampling not only of what appears to be wrong, but also of what appears to be right. This is not only to make available a 'control' for purposes of comparison, but also to permit detailed examination to check that what appears on the surface as visual evidence of good practice etc. is actually so. For example, a movement joint may appear from visual inspection of the surface jointing compound to have been correctly executed. Opening up may reveal that the joint extends in depth no more than the thickness of the top seal. Some years ago a contractor was paid for hundreds of metres of sewer 'measured' by walking from manhole to manhole. Subsequent investigation after the determination of the contract and bankruptcy of the contractor revealed that the manholes were set on solid ground and the sewers to which they were supposed to give access did not exist! An exceptional example of a common need to examine thoroughly and take nothing for granted.

Other experts at site investigations

If opposing experts are present, they may request an opportunity to make checks which for some reason or another the plaintiff's expert has not made. Recently a plaintiff's expert took moisture readings of roof timbers as part of an investigation into alleged roof leaks. Readings indicating approximately 12% moisture content were obtained, and the expert's reaction was a clear indication that this supported his contention, whereas the defendant's expert had other ideas.

At the request of one of the experts the moisture content of timber furniture was checked with the same instrument as used in the roof, and the readings in the furniture turned out to be significantly higher than those in the roof timbers! A good example of the value of checking everything as well as the desirability of taking control measurements. Certainly, other parties' experts must, whenever possible, be given the opportunity to make their own observations, make records and take samples where appropriate.

It is outside the scope of this book to demonstrate how a site investigation should be carried out or how and what records should be made. Sufficient to say that procedures will vary according to the type of failure and many other circumstances, and the investigator will normally prepare a checklist of activities, equipment and assistance needed on site, as well as such arrangements as are needed to ensure the presence of other parties where necessary and suitable access, etc.

Examination of relevant documents

Relevant documents may include any combination of the following:

(1) Correspondence between the various parties from original brief for the building to the point where legal action is contemplated (and often beyond).

(2) Contractor's contracts and/or agreements with sub-contractors and suppliers.
(3) Contracts and/or agreements entered into between designers and client, client and contractor and possibly sub-contractors and suppliers.
(4) Designers' drawings, specifications, instructions and bills of quantities.
(5) Reports by specialists including meteorological reports and/or records.
(6) Relevant legislation such as building regulations, etc., British and other Standard specifications and codes of practice.
(7) Textbooks, technical articles, records of papers delivered at seminars and conferences.

Documents not available in the early stages of investigation

At the initial stages of an investigation many of the above will not be looked at because they are not available and many others because their relevance is not established until much further down the line.

Examining drawings and specifications

Sometimes, even where original design drawings and specifications are available in the early stages, an expert will prefer not to look at them until he has examined the building in great detail. This is to ensure that his investigations are not influenced by a perusal of documents which could encourage dangerous preconceptions.

When they are examined it will be for a number of purposes. These will include comparing what was called for with what was actually built, checking drawings, specifications and bills of quantities for compatibility with each other as well as for compliance with standard codes and specifications and general good and sensible practice. The last of these will include 'buildability'. There are many details which appear sound, but can only be achieved by, to give an admittedly absurd example, erecting the paint and attaching the timber to it! It is also important to check dates at which various documents were prepared and the dates of any revisions. Sometimes a comparison of dates of one document with another may produce vital evidence. For example, the culpability of, say, a sub-contractor may well hinge on whether certain information was available to him when he did his work. Failure to check revisions and the date they were made could result in the expert receiving a completely false impression which might only be revealed to him when it is too late.

Investigating without the benefit of contract documents

Sometimes initial investigations are carried out on a building 'sold on' by the developer, or the owner is not in possession of detail drawings, and litigation is not contemplated. The expert nevertheless will have in mind the possibility of a dispute. He will therefore probably advise his client not to approach the original designer for assistance to make drawings etc. available until legal advice has been obtained. In such circumstances considerable investigation may be carried out without any of the original design information being available. Sometimes it is possible to view plans deposited with the local authority, but these are normally only for perusal in the offices of the authority, and also are seldom of large enough scale to be of great help. Notwithstanding that, they may be of some use.

State of the art

In comparing what was called for and what was built with good practice as evidenced by codes of practice, textbooks etc., from the technical angle, either it works or it does not. In assessing responsibility it is essential to check design and workmanship with the requirements of good practice as they existed at the appropriate time. Even this apparently simple exercise may be compounded by a change in standards, codes or working practices occurring during the design period: another vital reason for checking the dates at which documents were produced, revised and issued.

Laboratory examination

Scientific examination and testing of components and individual materials taken from a building may prove of immense value in establishing, confirming or proving causes or part causes of failure. It is very important to take great care with the manner in which samples are taken. There is a need to be aware of why the samples are to be tested and have a knowledge of the test procedure before they are removed from their built environment. They should be removed either by the person who is to conduct the tests or at least with the benefit of his guidance. The recording, for example, of moisture content, weather conditions, etc., may all be critical. Actual methods of removal and packing and transport may also be of great importance. So too may be the quantity of samples analysed or tested. In addition of course, it is sometimes desirable to test components or materials in situ. Such tests may go beyond the activities of checking moisture content of materials and temperatures and humidities within a building and comparing with external conditions. Where condensation might be a cause of problems, such checks are virtually routine. All in all, the owner with a failure on his hands would be well advised to anticipate the need for laboratory testing to a greater or lesser extent for any situation but the simplest.

On-going site inspection

Very often, if the timing related to the finalisation of reports and subsequent trial permits, much investigation and recording of evidence may be carried out during the progress of remedial work. This process however must be controlled in a number of ways for a number of reasons. There will be a need to establish before remedial work documents are finalised the scope, method, frequency and duration of further site investigation and recording work. The decisions arrived at must be made known to prospective tenderers for the remedial work via, and included in, the specification and/or bills of quantities. This is to ensure that the likely nuisance to the remedial works contractor is known to him in advance. He must be able to put a price on the extra administrative and possibly labour costs that will be incurred so that these additional costs are included in the contract sum rather than appearing as a subsequent claim by the contractor, perhaps after the case has been heard. Apart from visits and on site activities of the plaintiff's expert, there must be an opportunity, whenever possible, for defence experts to see if they wish (and they almost certainly will) original work as it is uncovered. They will almost certainly wish too to monitor the additional investigations and records being made by the plaintiff's expert as well as the actual remedial work itself. They may also wish to carry out investigations of their own. The extra cost of such work must obviously be kept to the minimum consistent with providing sufficient evidence to the court. Also, during the progress of the work significant defects may be uncovered which could reinforce the claims already made and/or give rise to new ones. All parties will need to see such work, and it could result in delays to the remedial works contract which would ultimately have to be paid for by either plaintiff or the defendants.

Facilities for defendants' experts

The type, scope, extent and sequence of site investigations will vary considerably according to overall circumstances. These may well involve discussion between members of the whole of the plaintiff's team. Wherever possible, arrangements will also take into account the reasonable requirements of the one or more defendants involved. This is partly as a matter of courtesy. It is also important to avoid any suggestion that the defendants' experts have not been given 'a fair crack of the whip'.

The information obtained during investigations on site, in the laboratory and from documents will in the majority of cases provide much of the plaintiff's technical case, and the technical defence. All this material requires detailed analysis which is dealt with in depth in Chapter 5.

Chapter 5

Post-investigation analysis

Facts and opinions

The job of the expert is to give opinions based on facts. There is a view that who gathers the latter is immaterial. This is completely erroneous. Chapter 4 touched upon the why and how of collecting the necessary information. Apart from now enlarging on both of those themes it is timely to discuss as part of the processes the importance of knowing which facts may be relevant and emphasising those that are. It is also important if the non-relevant are, before discard, explained as such if there is a probability of other people attempting to make capital of them.

It may not, of course, always be appreciated that other parties might consider important some fact that has been dismissed completely by one expert as being of no consequence. If it comes up at a late stage in the overall proceedings, the expert who has dismissed it must at least retain enough in his memory to be able to say why he considered it irrelevant. If he cannot remember, he may not always have the opportunity to explain his thinking. Should this be the case he may find himself in difficulty.

While the above paragraph is anticipating the later stages of the expert's work, the possibility must be borne in mind at the post-investigation stage when available information is being analysed in greater depth than during investigations.

Availability of information

As already explained, the information from which the expert will form his final conclusions may become available at various times and in various sequences. For this reason the different types of information are dealt with as separate entities, with an indication of the basic rules used to put all together in order to reach, wherever possible, positive opinions on the technical reasons for failure and the apportionment of responsibility for that failure, as well as recommendations for further action.

Laboratory testing and analysis

It is assumed that site investigations have been completed at least to a stage where a reasonably representative sample of defects has been investigated and associated investigatory work done, and a reasonable number of samples taken in the correct manner for the tests that are to be carried out on them. For example, British Standard specifications published by the British Standards Institution (BSI) lay down detailed requirements for a number of tests used to determine whether or not materials conform to a particular and relevant standard. In other instances it may not be BSI tests that are involved, but tests which still need the correct amount of material to enable the laboratory to work correctly. In other instances, testing for moisture content for example, there will be other needs. A material may be tested for moisture content in situ immediately on opening up perhaps, and then as soon as possible placed in a suitable container, which is then sealed for transmission to the laboratory in order that the moisture content may be checked by more scientific means than a site instrument. Sometimes paint samples need to be scraped from a surface and placed in a clean and safe container for subsequent checking to determine the number of coats and chemical composition of each. This may be necessary both to help establish cause of failure and to check for compliance with the designer's or manufacturer's specification.

It is also sometimes necessary for laboratory staff to carry out tests on site. For example, there may appear to be leakage between a window frame and the wall into which it is built. There may well be a need to check the amount of water that penetrates under different conditions of volume of water and air pressures both inside and outside the building. This would require quite a sophisticated test rig operated by personnel familiar with the appropriate testing criteria and able to relate the results to established standards in terms of exposure rating, i.e. the forces the assembly should be able to withstand relative to those pertaining on site.

Laboratory report to the expert

When the staff of a testing house have concluded their own investigations they will present their specialist report to the general expert. Normally such a report will give details of the samples, how obtained or received, and the nature of the tests. It will then go on to compare the results obtained with any relevant standard tests. This will usually be followed by conclusions which will be an opinion concerning the likely consequences related to the material under scrutiny. It is the job of the general expert to look at the conclusions in the light of all the other information in his possession and slot it into the overall framework of facts. This fundamental operation of analysing all the facts related to each other applies virtually throughout the diagnostic process, whether the diagnosis is technical or concerned with responsibilities. In fact in the majority of building failure investigations the two are inextricably interwoven.

Relevant legislation

There will always be a need to check what has been drawn, specified, measured by the quantity surveyor and actually built against the requirements of building legislation. Other legislation may also be involved, such as the Health and Safety at Work Act. There are frequently others, for example a cinema will have to comply with the Cinematograph Act of 1904, a public house with the regulations imposed by the Licensing Justices. This is in addition to fulfilling the requirements of the Building Regulations or London Building Acts and any planning restrictions. It is very important to remember that the relevant legislation (or regulations) is that which was in force at the time the project was designed and built. Sometimes regulations change during the design or build process, and in some cases the precise date of change may be critical in determining whether or not a designer or contractor should have complied with the new or the old. This is a situation that may also apply to trade literature, British Standard specifications and codes of practice and trade codes.

British Standards Institution and other specifications and codes

British Standards are specifications and codes of practice (the latter still titled as specifications) produced under the auspices of BSI. Those related to the manufacture of materials and components for building as well as parts of assembled buildings have considerable authority in the building industry. In consequence they carry much weight in the courts dealing with building disputes. Also in common use in the UK are European and other foreign standards which may sometimes be used in preference to British. These include DIN (German), ISO (International Standards Organisation) and occasionally American and Canadian standards may be specified.

The preparation of British and trade standards

These standards are normally drafted, or at least ratified, by a committee formed from members of the specialist industry relevant to the subject plus representatives of consumer groups and other interested parties. For example, the code of practice for the polyurethane foam insulation of cavity walls was prepared by a committee representative of manufacturers of the constituents of the foam, installers and architects and surveyors from both the public and private sectors. In addition there were representatives from the Agrément Board (the function of which is explained in the paragraphs which follow), the Building Research Establishment and other interested organisations.

British Standard and trade codes of practice and specifications are often very closely allied in content. Frequently the trade documents are prepared by many of the people who are also members of the BSI committees and panels. Sometimes they are prepared at much the same time, but almost invariably both the BSI and trade documents will follow, where applicable, established good practice. One point sometimes overlooked by designers who ask in their own specifications for compliance with British Standards is this: many of the standards have a number of options which have to be selected to suit particular circumstances. There is an often used specification clause which says something akin to 'All such and such to be in accordance with BSS XYZ'. It is possible for contractors, wittingly or otherwise, to comply with the standard but still produce work unsatisfactory in the context of the relevant situation through failing to observe the precise requirements detailed in the specification to cover a particular application. Sometimes the design will be at complete variance with a standard, and while the contractor has a duty to spot this and point it out to the designer, it does not always happen!

Agrément certificates

The Agrément Board in the UK was formed some years ago on similar lines to an established French organisation. Essentially, the board tests the performance of materials and components submitted to it against the claims of the manufacturer. The board issues certificates describing the properties of the products and these certificates lay down the manner in which the materials should be used if the stated possible performance is to be achieved. Thus in the manner appropriate to British Standards, it is important for specifiers to ensure that they know how a product should be used rather than saying to themselves 'It must be alright, it has an Agrément certificate' and then ignoring the guidelines laid down for satisfactory use.

Trade literature and textbooks

Much trade literature is, not surprisingly, designed with the principal object of assisting manufacturers to increase their sales. However, there is also some trade literature produced perhaps for similar motives, which presents information in a manner which is of considerable value to designers and/or contractors. In this respect it both assists them to use the products correctly and also presents in detail the whole assembly or assemblies of which the product in question is a part. While there may be a commercial bias, the literature is still in the nature of the sort of information that may be found in textbooks. Other trade literature, and, indeed, textbooks, will be geared to producing information for the tradesman who is to install the product and also the user who has to maintain it. All these types of information need checking for compliance by designer, contractor and user.

Meteorological records

Many instances and types of building failure are influenced by weather conditions, be they rain, wind, temperature, humidity or a combination of two or more of them. Reference to meteorological records can often be of considerable assistance in making or confirming a diagnosis. Such records need to be interpreted with due regard to other information. For example, take the case of water dripping from the under surface of a top-floor ceiling. It may be argued that if drips appear during fine weather the cause is something other than a roof leak, perhaps a defective pipe joint or even condensation. In fact however, an appraisal of all the facts may reveal that there is a probability of substantial delay between rain penetrating a roof membrane and the water reaching the underside of the total roof assembly, so that rainfall may be a contributor to or even the sole cause of the problem.

Assessing the facts

The expert will be starting his analysis armed with some considerable knowledge of his subject. He will not of course be completely familiar with every aspect and every detail of building construction and practice. Neither will he have at his fingertips full awareness of all relevant legislation. Nevertheless, he must have the experience to enable him to appreciate what he needs to find out and how and where to do so. His analyses of why things went wrong and how to put them right will be based on the most up to date information available. His final analysis of responsibility must however be based on the most up-to-date information available at the time of design and construction. This is a factor sometimes overlooked, but is of the utmost importance when considering what should have been done at any particular time. If this 'state of the art' changed during the design and build process, an assessment of what should have been done can become very difficult. For example, a new British standard code of practice could be promulgated towards the end of the design period of a particular project. There is a need to decide whether or not the designer should have known about it. If he should have known, should he have incorporated it? If extra expense and/or delay would have been involved, should he have explained to his client the likely consequences of the various options and asked for a decision?

This sort of situation can only be decided on its merits and there can be no hard and fast rules. Also in terms of assessing responsibilities it is important to know the duties of the respective parties. However, this is putting the cart before the horse. The first essential following an investigation is to ensure beyond reasonable doubt that *all* the symptoms have been established. It is equally important to ensure that there are no apparent symptoms which are not in fact coincidental red herrings. The second step is to examine carefully those symptoms in the light of all the available factual information discussed earlier in this chapter and to compare what is shown on the design drawings and specifications with what has actually been built, and then to compare both design and construction with the expert's knowledge based on training and experience and with the various types of information discussed earlier. That is to say, that which is found in legislation, textbooks, standards and codes and factual records, including test results, as well as what is known by the expert to be successful common practice in the building industry.

Once this stage has been reached there are a number of possibilities. The cause(s) of failure will be:

(1) Obvious and irrefutable.
(2) A matter of opinion based on a logical analysis of all the facts.
(3) A mixture of (1) and (2).

In an ideal world the three major possibilities would be recognised by all parties' experts and this book could finish after a brief discourse on remedial work. Unfortunately, experts not only do not always share opinions, but frequently have differing opinions on the subject of logic. Sometimes some of them apparently fail to understand logic anyway! Thus, there are often profound differences of conviction on the interpretation of fact into opinion, and sometimes on what constitutes fact! Even the irrefutable in terms of technical issues can become very much a matter of attitude when responsibilities are under discussion. Complete objectivity, however desirable, and in the long run sensible, is not always achieved by all experts.

However, at the stage of analysis by the plaintiff's expert, it will be only his own judgment (in the non legal sense) with which he needs to concern himself.

Assessing responsibilities

Once having diagnosed causes of failure, the expert will, in the majority of cases, be asked at some stage or stages to advise on remedial works and on responsibilities. The former may be split if necessary into emergency and permanent. In both cases some indication of cost will almost certainly be wanted by the building owner and some indication of the likelihood of being able to recover a substantial part of the cost of both work and fees from others. Estimating costs will very probably require, except in the simplest of cases, input from a professional estimator such as a quantity surveyor.

By the same token rarely will the expert be able to venture an opinion on recovering cost without the assistance of the lawyers. He may be able to give a very sound opinion on responsibilities without assistance, but the chances of obtaining satisfactory financial redress by litigation or arbitration are another matter.

In assessing responsibilities, there are other than purely technical matters to be taken into account. It is, in the majority of building disputes, essential for the expert to have a detailed knowledge of the duties of the various parties as provided for by the various contracts and agreements into which they enter. There will be, for example, the contractual relationship between the building owner and his designers and specialist consultants as well as the interrelationship between the consultants themselves. In the case of a building designed predominantly for human activity, the head of the design team may well be an architect with structural and services engineers employed directly by the client but with their activities co-ordinated by the architect. In the case of, say, a power station, the plant engineers may well be the team leaders, with the structural engineer in a slightly subordinate role and the architect even more so. Some projects may be primarily structural with services engineers and architects in a subsidiary role. Frequently, particularly where structural and services inputs are limited in scope, such consultants may be employed directly by the architect. This, however, is a situation not much loved by those who insure designers against claims of negligence. Once a project is about to become a physical reality there are whole new sets of contractual relationships. That existing between employer and contractor, and possibly others between employer and sub-contractors and between main and sub-contractors. Involved with all of them are the professional consultants and resident site supervisory staff.

Agreements between the parties

The expert will need to be aware of the provisions of all the relevant standard forms of agreement. These will include of course agreements between client and consultants. In the case of 'design and build' schemes there will be situations where private professional consultants are working for and directly employed by a firm of contractors. The expert will need to be aware not only of the forms of agreement but also of their practical effects. This means a thorough grasp of the duties of all the parties involved in the entire process of designing and building from initial approach of client to designer (with the recent relaxations in codes of conduct, perhaps designer to potential client) to settlement of final account between client and contractor. As for technical information, it is not suggested that the expert carries all in his head, but he will be aware of the principles, know where to find the detail and be able to interpret it.

Remedial work considerations

The expert will have the know-how to assess the structure he has investigated in order to establish the best ways of designing and implementing remedial work. He will need to weigh up the technical issues to ensure that the remedial works are successful and be sure that in solving one problem he does not create another. He will need to be able to time the execution of the remedial work with an eye to both the technical and legal consequences of his actions. (Why is explained later.) He must design remedial works to be as economical as is consistent with efficiency, and be able not only to distinguish between essential work and 'betterment' (desirable but non-essential improvement) but able to present the differences in a form understandable by and acceptable to a court.

Summary of immediate post-investigation analysis

To summarise the essentials of the early post investigation analysis:

(1) Ensure all symptoms are investigated and the facts assessed logically to arrive at the opinions.
(2) Ensure *all* causes of failure are identified.
(3) Compare causes of failure with good design, workmanship and maintenance practice.
(4) Compare causes of failure with good procedural and contractual practice.
(5) Check what each member of the design and build team did against what they contracted to do.
(6) From the above, form and give his own opinions regarding responsibilities.
(7) Design and then check remedial proposals for technical effectiveness and appropriate procedures and timing.
(8) Ensure that all his conclusions will stand up to detailed scrutiny.

Then – check everything!

Information retrieval

It will be remembered that the documentation used by the expert falls into two distinct categories. First, there are the drawings, specifications, minutes of meetings, bills of quantities, correspondence and similar documentation which relates to the particular project. Second, there is information that may or may not be mentioned in the first category but is available for general use throughout the building industry.

Taking the first, documents initiated for the particular project will need, of course, to be carefully filed and indexed in order that the expert may readily lay his hands on any of them quickly and easily. The second category, many of which will be in his library, will also need to be readily available for reference. Most experts build up a library of out-of-date information. This is to facilitate basing conclusions on the 'state of the art' as it existed at the relevant time of design and construction. He will probably not have in his office all the older documents needed for reference, but will know from where to obtain them. All in all, a simple but efficient system of filing and retrieving both categories of information is essential.

Chapter 6

The report of the building owner's expert

It could be considered that completion of the work summarised at the end of the previous chapter means the report has virtually been written. Certainly a great deal of the preparatory work which precedes the writing of the actual report has indeed been done. Turning that into a document which will put the owner's case as accurately, succinctly and compellingly as possible demands, however, a great deal more thought and effort.

Privilege

Before discussing the content, order and format of the report there is an important legal point which should be borne in mind. Documents which are created for the dominant purpose of litigation may be held to be privileged. This means that they do not have to be divulged to opposing parties. An expert's report produced for the purpose mentioned would fall into this category.

It must not be assumed however, that all experts' reports or drafts of reports are privileged from disclosure. There is frequently doubt as to whether or not reports prepared by an expert are privileged. This is particularly applicable to reports prepared early in an investigation. Their predominant purpose is as likely to be considered the diagnosis of the faults and subsequent rectification as it is to be litigation. Furthermore, documents which are initially privileged in themselves may cease to be if referred to in documents which are disclosed to the other parties. At the very least, such a reference is likely to produce a request for a copy from an opposing party, with possibly an application to the court should the request be refused. Even if the application is unsuccessful, time and money will have been spent by the client.

Very often the initial report of the expert may be written before all the evidence, both technical and other, has been obtained. In consequence, the conclusions contained in earlier reports may differ from those expressed subsequently. Given the considerations mentioned in the previous paragraph, it is advisable for any report to be written with the possibility in mind of disclosure to other parties in the event of litigation. Where, therefore, opinions are qualified, the reasons for qualification should be given, and no opinion stated as definite unless the author is satisfied that no circumstance will arise that could result in a change being made that would require subsequent explanation. When this is not done, a change of opinion might appear, quite erroneously, to be a belated attempt to present a client's case in an unjustifiably more favourable light, with a resulting loss of confidence in the expert by the court.

Privilege can be a very complicated matter and, as far as the expert is concerned, problems are avoided if disclosure of all reports is assumed possible. It is worth identifying the status or edition of reports only by the date of completion and not by a description such as 'Final Report' which draws the attention of the reader to the existence of earlier reports which, quite legitimately, may not have been disclosed.

The text of the report should be selfcontained and avoid reference to earlier reports to avoid any possibility of a request for disclosure. Even if an earlier report has been disclosed it will avoid the need for the reader to refer to another document. In addition, the use of a date only to identify the report edition will avoid ambiguity when referring to it in conversation or correspondence.

(*Note*: The rest of this chapter uses an example of a numbering system for sections, clauses and sub-clauses often found in an expert's report. The purpose of this and similar systems is to make it easy to identify each part of a report without the need to renumber every paragraph if amendments are made in the form of additions, deletions or rearrangement during drafting.)

1.0   Content of Report

Like any other document, the first requirement is to appreciate why it is being written. It is incredible how often this fundamental point is overlooked by some self styled 'experts'. The 'why' includes for whose consumption it is intended, what it is trying to tell them and for what purpose. The initial report prepared for a building owner and perhaps for his legal advisers may well be quite unsuitable for future readers and users. It may well suggest that there should not be any future readers. On the other hand, if the expert has discovered faults of design and/or workmanship or materials, it could finish up as a major part of a building owner's case for redress of an alleged wrong. It must be clearly understood that the previous sentences do not mean changing opinions to suit the arguments or convictions of one's client, or a client's lawyer! It may, however, mean that the opinions of the expert change as further evidence comes to light. If the preparatory work discussed in Chapter 5 has been properly done it is unlikely that technical views will alter substantially, although it can happen on occasion. However, there are numerous instances where a preliminary indication of cause of failure and responsibility is called for long before all the information for complete diagnosis is to hand. Where this happens the expert will make clear in writing that his findings are as close as he can get to an opinion within the limits of the information available to him at the time.

1.1 Common patterns in reports

There are no hard and fast rules for deciding what should go into a report any more than there are for deciding format and order. Nevertheless, there are likenesses between a large number of disputes in the building industry that have resulted in an element of similarity in the basics of presentation of experts' reports. Going back to the elementary question of 'why' it is fair to say that the majority of initial briefs to an owner's experts say in essence: 'There is a problem with my building which is causing me worry, aggravation and loss. Please tell me what has caused the problem, how to put it right, how much it will cost and whose fault it is that the problem exists'.

1.2 The brief in detail

To develop the outline presented in Chapter 3, it is usual and advisable for the expert's brief to be set out in some detail very early in the report. It is then clear to all reading the report precisely what has been called for. It is also clear to the expert himself precisely what he has been asked to do. The actual process of setting out and checking the objectives will often highlight to the expert that his brief is not in fact correct. It may be inadequate, and further instruction or authority needed to cover vital aspects for which the expert has no mandate. Once it has been correctly set down and agreed with the building owner it serves as a checklist to ensure that when the report appears to be complete, it actually is.

Because the expert is briefed in successive stages, each edition of the report will contain the new briefing. The brief as it appears in the report will probably vary from the original instruction from owner to expert and will be revised as interim reports, either oral or written (perhaps just in letter form), are made by the expert and considered by his client and, if relevant, by the client's lawyers. This will of course depend very much on the opinions of the expert as his investigations progress.

1.2.1 *Background or introduction*

It is desirable for the initial brief to be augmented by a short history of the building, including its age, the names of the designers and contractors, and any other relevant organisations, and the reasons for the appointment of the expert. This may, depending upon circumstances, either precede or follow the statement of the brief.

1.2.2 *Symptoms*

Almost certainly the symptoms of failure that initially prompted the brief will have already been mentioned. Nevertheless, it can be useful to conclude the setting out of the owner's brief with a restatement of those symptoms which will firmly set down the need for the report to have been prepared at all.

1.3   Available documentation

It is usual for the expert to detail the information available to him during the preparation of the report and this will involve some, but not necessarily all, of the following:

(a) Designers' and sub-contractors' drawings and specifications.
(b) Instructions issued by the designer during the progress of the work.
(c) Bills of Quantities.
(d) Minutes of meetings.
(e) Correspondence files. (Designer/client, designer/contractor, contractor/sub-contractors and suppliers, etc.)
(f) Clerk of works' reports and diaries.
(g) Relevant building regulations and other legislation.
(h) Technical literature, British Standard specifications and codes of practice.
(i) Manufacturers' instructions for use of proprietary materials.

(j) Agrément certificates.
(k) Specialist experts' reports.

The above list is reasonably, but not completely, comprehensive.

When the final version of a report is produced not all the documents mentioned above may be available. This applies particularly of course to defendant's (respondent in arbitration) documents which may not be disclosed (when each side's documents are made available for inspection by all other parties) in time for inclusion. When this happens there may be a need for a supplementary report to be prepared. It is usual somewhere in a report, often at the end of the list of available documents, to mention that revision or a supplementary report containing revised opinions may be necessary in the light of subsequently produced documents. Also, where reference is made to codes of practice, standards, catalogues and similar documents it is essential to give the date of the document relied on to demonstrate that it represented the 'state of the art' at the appropriate time. It is also usual for documents relied upon to be readily to hand for readers of the report, often in the form of appendices.

1.4   Site investigations

The expert's notes on site investigations will normally record all visits and investigations made, including the names of other persons attending and the capacity in which each was present. It is normal also, unless the nature of the problems renders it unnecessary, to record weather conditions during investigations. Indeed, a record should be made as required by circumstances of any relevant information. The inclusion in the report of any of the information recorded will depend on the nature of the individual record and the circumstances of the case as a whole.

1.4.1 Opening up

All opening up work should be recorded, including the sequence of operations, who did the opening up, who observed it, together with details of what was found at each stage. Opening up should only be done under the strict control of the experts. If records are referred to in the report their location should be cross-referenced. This cross-referencing may take the form of a margin note or a note in brackets of the relevant appendix or other location of the information. At this stage of the report, however, descriptions of site investigations, whether written, drawn or supplemented by photographs, should be just that – descriptions. They should be comprehensive and detailed unless subsequently discounted as irrelevant. Even then, they will need to be mentioned briefly elsewhere in order to explain why they are irrelevant.

1.5   Discussion of all information

This section weighs up the significance of what is found on site and relates it to *all* the information available at the time the report is drafted. It will normally analyse the design and construction and tie in both to the symptoms of failure. In doing so a picture will begin to emerge of the physical causes of failure. As the expert develops his clarification of physical causes, he will start to relate these to standards of design, construction and maintenance as well as to natural phenomena. How he expresses these thoughts in relation to the chapters which follow is a matter to be decided by circumstances, experience and individual practice.

1.6   Technical conclusions

Depending on whether or not causation is already mentioned, this section will either start with and then develop the causation theme or be a development only. Either way it will be a logical progression through all the reasons for the failure(s), the principal through to the minor, but dealing entirely with the technical reasons, be they one of design, construction, materials, maintenance, natural phenomena or any combination of two or more. Within each of those five groups there may be several factors. The overall diagnosis may, and frequently does, prove to be the consequence of a whole series of events.

Perhaps it is this section of a report that needs more than any other to be a model of succinct, clear, simple, jargon free expression with the logical progression mentioned above. It is here that the layman is going to be either enlightened or confused. Remember, in terms of technical conclusions, the building owner, solicitor, barrister and the judge, while very probably possessing a fair degree of building knowledge, are laymen when compared with the expert. Remember also that one clear diagram, sketch, photograph or model may be worth a thousand words.

1.7   Conclusions related to responsibilities

An expert is nearly always asked to say, as part of his brief, who, as well as what, is in his opinion responsible for the failure or series of failures. The expert must remember that in legal terms it is he who is the layman. An expression of opinion on this aspect of failure must be made with considerable care to ensure that what he says is strictly within his sphere of competence.

If as a result of what has gone before he is able, for example, to say that a failure has arisen as a result of faulty design this does not necessarily mean that the liability lies with the designer. The designer may have been working in accordance with established practice of the time which is subsequently proven incorrect. He may have been working, under protest, within the requirements of his client. He may have designed with the assistance of specialists who were at fault with the advice they gave, advice beyond the knowledge he could reasonably be expected to have.

A failure could be the consequence of incorrect or inadequate maintenance, but responsibility lie with the manufacturer or designer who gave the wrong maintenance instructions. Poor workmanship could be of a nature that should have been seen by the designer and his site supervisory staff. In such an instance some responsibility may lie with the designer, depending on the nature of the work. For example, if repetitious work was consistently poor some blame would probably lie with the inspecting designer, while a 'one off' might legitimately not have been seen if executed between visits to the part of the work in question. It is for reasons such as these, and many others, that the expert needs to see as much documentation as possible. In addition, however, he must have a first-class knowledge of the duties of all the disciplines involved in the design, build and maintenance processes. He must also have the same degree of expertise with regard to the interaction of the roles of each of the various parties.

When the expert has considered his technical conclusions in the light of all the information in his possession he can begin to assess responsibilities. To do so he must equate what each party did to what he should have done in contractual terms, and check both against the requirements of good practice, in order to form an objective opinion as to whether that party acted with the reasonable skill and care to be expected of a member of the discipline in question. He must discount any specialised knowledge possessed by himself as an expert that the defendant, be he architect, engineer, contractor or whatever, would not be expected to have.

Sometimes the expert may be given the difficult task of assessing the degree of responsibility of each party. If, as often happens, responsibility lies with more than one party to a dispute it is virtually impossible to reach a completely accurate and thoroughly objective conclusion. It is unlikely that the expert will be asked to include an estimate of percentage degrees of responsibility in his actual report, but he may be asked to give a separate oral or written opinion based on his knowledge of the case.

It is important how the expert sets out his opinions related to responsibilities. These should state what was done or omitted, what should have been done or omitted and, in consequence, the way in which he considers those involved failed to fulfil their obligations. It is the prerogative of the court to decide whether or not a person has been negligent. Realise incidentally, that prior to the end of a case it is only allegations that are being made against anybody. This is a point that can take on new meaning when the building owner as plaintiff sees the defences of the other parties, not to mention possible counterclaims!

1.8 Recommendations for further action

In preliminary reports, recommendations for further action, or perhaps non-action, are likely to fall into the following categories:

(a) Further investigation, including possible investigation by experts of other disciplines.
(b) Remedial work.
(c) Litigation or arbitration.

Taking these in order:

1.8.1 Further investigation

A recommendation for further investigation is quite common in a preliminary report, but, of course, less likely in later ones. It will normally only be required where the original brief to the expert has been limited in scope. In such circumstances the expert may have no authority to do everything necessary to confirm a diagnosis beyond doubt, or even to reach tentative conclusions on the causes of failure. Sometimes the expert may be satisfied in his own mind of the causes but be aware that investigation by other experts may be essential to convince a court. For example, an architect expert may have a fair degree of experience related to double glazing failures, but the lawyers and the expert himself may feel the need for the additional weight that investigation by a glazing specialist would lend to an argument. Further investigation/recording of evidence is also common during the progress of remedial work and, if not already cleared, will require the sanction of the client.

1.8.2 *Remedial work*

The stage at which remedial works are considered in relation to both emergency holding repairs and permanent work will depend on a number of factors. If water is pouring into a building, the expert's initial report may well be an oral one, followed or even preceded by instructions to a contractor to carry out emergency repairs with no thought other than stopping the flow of water! If on the other hand, desperate measures are not called for, a number of matters will influence the course of action. These may well include cost effectiveness, disruption to use of building and potential future legal action. The failure may involve temporary safety measures such as 'fans' around the outside of a building to protect persons and property from possible danger due to falling external finishes. In such a case (loose external wall tiling is an obvious example) permanent remedial work may be, from the purely technical viewpoint, left in abeyance for several months provided safety is assured and there is no danger of, for example, serious deterioration of the fabric.

The expert must treat each case on its merits and make the appropriate recommendations to the owner at the appropriate time. Generally, it will be necessary to advise on the timing and nature of both temporary and permanent remedial work incorporating, among other matters, the following:

(a) Any initial emergency action, including design, implementation, costs for work and fees.
(b) Retention of temporary scaffolding, fans or similar until permanent work done.

(c) Design and contractual methods for permanent work.
(d) Estimated cost and fees.
(e) Timing related to detailed investigations, testing, design, observation by other parties and legal advice.
(f) Timing and extent of both temporary and permanent work related to 'mitigation of loss'.

The phrase 'mitigation of loss' expresses the principle which requires a building owner to carry out remedial work at a time and in a manner that reduces cost of work and other costs to a total that is a practicable and reasonable minimum.

The principle needs some elaboration at this stage. Although doing extra work within the same contract as essential remedial work may be a sensible measure, the defendants, if found generally liable, cannot be expected to pay for work carried out for the benefit of the owner, but unrelated to the failures. For example, say essential remedial work requires the scaffolding of a brick-clad building. The owner may decide to take the opportunity to have the brickwork repointed. The work is not the responsibility of the defendant and would be classed by the court as 'betterment' to be paid for by the owner/plaintiff. But who pays for the scaffolding? A question for the lawyers to answer, not the expert, although his opinion may well be sought on whether the scaffold needed extra work or a longer hire period than would be the case if repointing were not done at the same time.

Another example of betterment would be if, during the course of essential remedial work, the owner decided to upgrade his building by the installation, say, of double glazing not required in the design brief for the original scheme.

It is also important that the quality and life expectancy of remedial work are related to what the owner should have been provided with initially. Equally, remedial work should not have a greater life expectancy than the remaining life of the building as a whole, unless any other course is impossible or, of course, it is asked for and paid for by the owner.

1.8.3 *Alternative recommendations for further action*

It may be that neither litigation nor arbitration are recommended. For example, the expert may consider that the faults that exist are entirely the responsibility of the owner. This could be as a consequence of poor or non existent maintenance or misuse of the building. Floors are designed to take a maximum safe load, and provided that load has been spelt out by the designer to the owner, the imposition of excessive loads may constitute misuse for which only the owner or his employees could be blamed.

1.8.4 *Seek legal advice*

Once the expert feels that there may be a case for someone else to answer, his normal recommendation would be that legal advice should be sought to confirm or refute his (the expert's) opinion that a technical case could stand up in court. At preliminary report stage some dialogue between technical and legal experts would be needed to arrive at a positive recommendation to the owner to litigate, arbitrate or neither.

Often the expert's initial brief comes via or at least after the involvement of the owner's solicitors, and dialogue between lawyer and owner is probable whatever the expert's recommendations.

1.9 Other contents

Most of the report content already covered in this chapter is in the order in which it may appear in a finished report. However, there are a number of other items usually included, located to suit the preference of the expert. These might, and normally will, include some or all of the following, and there may be others not referred to here.

(a) A curriculum vitae of the author, which will, of course, establish his qualifications and experience outlining his suitability as an expert.

(b) Copies of relevant documents referred to in the report, or extracts from those documents, including some or all of those named as documents available and used in the investigations. Precisely what are included will depend very much on the particular circumstances and on the normal practice of the expert, and probably as a result also of discussion with the lawyers when the final edition of the report is under consideration. Usually all of these will be presented as appendices to the main document. Whether or not main report and appendices are bound in separate volumes will depend almost always on the total quantity of paper. It is useful for the appendices to contain also a list of samples which are relied on, especially where such samples are to be produced at trial or arbitration. Essential in most cases are site and location plans of the building(s), including compass points.

2.0   Format of report

Despite there being no hard and fast rules for the format of a report, some basic guidelines can be given which are close to the standard practice of most established experts. A title page which forms or repeats the front cover would usually say for whom and why the report has been prepared, but in as few words as are essential to identify the subject. This title page would also normally give the name, address and telephone number of the author of the report or of the practice of which he is a part.

This would usually be followed by a contents page giving section headings and the page numbers at which each of them starts, as well as the title and location of the various appendices, e.g. name of document or documents, volume number or letter and page number. As far as the body of the report is concerned, the order would probably be much as the order of the preceding sections of this chapter, i.e., expert's brief, history and description of the project (including site plans), documents used, site investigation, discussion, conclusions (technical), conclusions (responsibilities), conclusions (remedial work) and recommendations for further action. Very often the main report finishes with short summaries of, if not every section, certainly the conclusions and recommendations. Some experts put a summary early in a report, but, provided there is a reasonable contents table at the beginning, it seems more logical to summarise at the end.

The above would be followed by the appendices, which would follow the order of the main report, except possibly for the authors cv etc., which may be early in the main report, or at the end of it, rather than as an appendix.

3.0   Presentation

The method of presentation is largely a matter of individual preference, but there are important points, some already referred to. Essentially these are the need for clear cross-referencing to facilitate use of the report both in court and elsewhere, page numbering, easy access to confirmation of facts (copies of documents in appendices) and a simple numbering system to facilitate identification of paragraphs. It is still common in many reports for main paragraphs to be numbered consecutively right through a document from one onwards. This can become both unwieldy and very time consuming if reports have to be amended (even with the use of word processors), and by numbering paragraphs and sub-paragraphs in a similar manner to that used in this chapter time can be saved in the typing and editing of reports as well as in conference with legal advisers and in court.

Several copies of a report, complete with appendices, may be needed, and such things as photographs need some attention to ensure easy identification of the correct negative or print when copies are needed. Colour photocopying is of a fairly high standard, but even when practising economy as far as possible, the use of reprints of photographs from negatives rather than the use of photocopies is to be recommended. A less than perfect copy can be dangerously misleading. By the same token, colour photographs are to be preferred to black and white.

Regarding the size of reports, there appears at times to be a school of thought among some experts that the larger and longer both the report itself and its appendices, the more likely it is to impress the court. In fact, a good report is as short and succinct as is consistent with telling the whole story.

4.0 Things to remember

There are a few points which need emphasising or restating with regard to the report of the building owner's expert when the former is also the claimant or potential claimant.

4.1 Loss

Any claim which has necessitated the involvement of an expert and the preparation of a report must be the consequence of loss. Even if bad practice of any sort is obvious – no loss, no claim.

4.2 Timing

The appointment of an expert as early as possible after the discovery of defects is usually essential rather than merely desirable.

4.3 Development of the report

The first report of the expert may well be oral or in the form of a short letter. Both expert's brief and information may extend as events unfold, and what starts as perhaps dealing with an emergency may end as a comprehensive report dealing with many aspects of the initial problem. The timespan from first involvement to completion of final report may be as long as several years.

4.4   Privilege

When reports are prepared, addressed and defined as to status or edition it should be remembered that it is always possible to waive privilege, but it is much more difficult to retrieve it if it has not been safeguarded in the first place.

4.5   Thoroughness

To be thorough is a vital attribute of the expert. When a report has been completed, at whatever stage, it should be checked and checked again to ensure that it fully meets the brief, explains all that needs to be explained, substantiates all that needs to be substantiated. Spelling, punctuation, numbering, references, etc. must all be thoroughly checked. However carefully checked, and however many times, it will still have some mistakes, but at least they will be minimised in both number and import.

4.6   Impartiality and objectivity

Whatever the initial circumstances, a report should always be prepared with the possibility in mind that it may be disclosed to opposing parties in litigation or arbitration and thus finish up as part of the agreed bundle of documents in a law suit. In consequence, the author must remember that he may well be cross examined with regard to the content. Also, long before the stage of disclosure to opposing parties, it is vital for his client and his client's lawyers to know exactly what the expert really believes. Woe betide the expert who allows himself to write his report or any part of it on any other basis. If he does, he is liable to do both his client and himself considerable disservice. It is acknowledged that these points are made elsewhere within these pages, but their importance cannot be overstressed.

Chapter 7

The way forward

Many of the further actions referred to briefly in the previous chapter are discussed in detail below. They follow much the same order as before but it must be accepted that no two cases are identical. Rigid rules cannot be laid down for the progression of any action, legal or otherwise. What may be the first event or activity in one case may, in another, be much later in the proceedings or perhaps not occur at all. Similarly, the relationships between different activities will vary in terms of type of activity and timing.

Further investigation

Take first the possibility, in an initial report, of a recommendation for further investigation. Such recommendation could be made for one or more of the following reasons.

(a) Technical issues are, at the end of the initial investigation not clear-cut in terms of causes of symptoms

This could require additional investigation by the original expert, perhaps because the initial investigation was limited due to lack of time, lack of finance or the requirement to consider the implications of what was originally found. The first of these could be brought about by a need to deal with an emergency situation, the second is obvious, the third is the consequence of looking at symptoms in the light of access to drawings, opening up of structure or the opportunity to consult textbooks or other authoritative sources. If finance is the problem there will be a requirement, at a very early date, for owner and experts to make very difficult judgments and weigh up the possible consequences of inaction. It is particularly difficult to establish guidelines for a situation likely to relate to small properties in single or near single ownership. Usually some form of financial assistance can be found and a degree of further investigation established to improve the chances of putting the property in good order within the constraints of available monies, from whatever source obtained.

(b) Initial investigation reveals problems beyond the competence of the original expert

Sometimes, perhaps, an architect expert will be aware that one of the problems in a building may have a structural cause that needs the input of a structural engineer. Consider for example a leaking flat roof. Visual inspection may reveal to the architect cracks in the asphalt membrane after removal of paving above the asphalt. Removal of suspended ceiling tiles may reveal cracks in the concrete roof deck, coincident or otherwise with the asphalt cracks. This may alert the architect to the likelihood of a structural problem with which he does not consider himself competent to deal. Examination by a structural engineer may well result in the engineer needing to undertake further investigation to establish beyond doubt the cause of the cracking of the concrete, and to establish whether or not such cracking is structurally dangerous. The architect may need to carry out further investigation to establish whether or not the cracking of the concrete was responsible for the cracking of the asphalt. It may not have been.

(c) Legal advisers require a greater weight of evidence

Sometimes, the original expert may be certain of the cause of a problem but such cause may be outside the normally accepted competence of the expert. Chapter 6 mentioned the possibility of a glazing failure established by an architect substantiated by a glazing expert preparing a report containing specialist evidence which the architect could have some difficulty in achieving on his own account. Similar situations might well prevail in connection with paint failure, rot in timber and many others.

(d) Original expert needs more help

At times, when causes of symptoms are not clear-cut, the services of different disciplines are needed by the original and principal expert in order to establish those causes. In some cases the need for specialist advice or even a second opinion from an expert of the same discipline may be apparent before much investigation has been made at all. Then, provided the employment has been sanctioned by the building owner, it may be possible for all necessary disciplines to do all that is needed in the course of one investigation. This is, however, the exception rather than the rule. Usually, it is only after an initial appraisal of the problems that the precise needs for assistance can be assessed.

Emergency remedial work

The execution of remedial work is another matter subject to a variety of variables. There may initially be emergency work, almost certainly of a temporary nature. In many cases it will be not be arranged by the eventual expert. Possibly by the building owner himself, probably by the original designer if known and available, or perhaps by a local architect, surveyor or engineer. Whichever, it is likely that such work will be carried out with the single objective of making the building safe/watertight and minimising further damage to the building and its contents, with little thought given to eventual permanent work or to the questions of establishing responsibility and recovering consequential costs. In consequence, there is a probability that valuable evidence may be unwittingly destroyed. It is nevertheless a situation difficult to avoid. Emergencies are emergencies, and the building owner who at a time of crisis is in a position to foresee the needs for the future is understandably rare. Even if he does foresee them, the immediate problems may make it impossible to do everything desirable in terms of subsequent investigation and action.

Permanent remedial work

Occasionally building owners (or their legal advisers) are reluctant to appoint an expert to design permanent remedial work on the grounds that the court will feel that he may not produce as economic a scheme as is possible. This argument may equally apply to a designer other than the expert. The real point is of course that the defendants will be trying very hard to establish as much 'betterment' as possible, and whoever the remedial work designer, he will know that his work will be subject to very close scrutiny. If it is clear that the original designer was in no way responsible for the failures, the building owner could very reasonably go back to him for the work.

There is, however, a potential problem with such a course, in that one of the defendants might join the original designer as a third party, with possibly embarrassing consequences. An alternative designer may well be a wiser choice.

The logical person to design the remedial work is really the expert. During his investigations he will have acquired a first-class knowledge of the construction and the problems to be overcome. However, there are experts who will not undertake remedial work under any circumstances. Sometimes they are too busy to do so.

It is vital nevertheless that if the expert is not to design the permanent remedial work, the building owner avoids the combined employment of expert and other designer in a way which could result in a situation of divided responsibility should the future performance of the building not be what it should.

Timing of remedial work and mitigation of loss

The courts require a plaintiff, as far as possible, to mitigate his loss. This means the cost of remedial work must, as suggested earlier, be kept as low as practicable. This not only requires an economic design scheme and the most satisfactory tendering procedures, but also means that the work must be timed to avoid extra work resulting from further deterioration due to delay. Inflation is also to be avoided if possible, as are costs due, where applicable, to disruption of the work normally carried out within the premises.

Nevertheless, the normal procedure is for the defendants to an action to have ample opportunity to appoint their own experts, and for those experts to have the chance to be present during investigations, to carry out their own investigations if they wish, and to have access to all evidence available to the plaintiff's expert(s).

Another factor regarding timing related to mitigation of loss is that only genuine lack of funds is normally accepted as a valid reason for delaying the implementation of remedial work. The time taken to establish the causes of failure and then design work to correct them without creating other problems would not, provided it is not excessive, be classed as delay. Also, it is accepted that once remedial work has been designed and approximately costed, there is still a period to be allowed for arranging finance, obtaining, checking and approving tenders and the signing of contract documents.

Nevertheless, it may at times be difficult to convince a court that there has not been unnecessary delay, and care should be taken to ensure that genuine delays can be substantiated.

Disruption during remedial work

Emergency work excepted, remedial work in terms of who designs it and when it is done needs careful consideration by the building owner, preferably in consultation with both legal and technical experts and, in many cases, financial experts as well. Also, whatever the building or buildings there will invariably be a need to take measures to reduce as far as possible disruption to the occupants and their lawful occupations.

Often, continuing with or rehousing the activities within the building(s) during the progress of remedial work can be, in percentage terms, a major item of financial loss. For example, disruption of production will almost certainly be a very serious problem in a factory. Nuisance to tenants, and in some cases temporary relocation, can be a costly matter for a local authority faced with major remedial work to a housing estate. Of course, determining precisely what costs may be legitimately added to a claim is a matter for the lawyers (perhaps assisted by accountants) rather than the technical expert, though he may well contribute much in an advisory capacity.

Action to recover costs

The other major factor mentioned as one of the recommendations for further action that the expert will make in his report is whether to proceed with litigation, arbitration or neither. For the moment, consider the first two as one, as the primary decision to be made is whether there are adequate grounds for successful action against some party other than the client.

The cases where there are no grounds for action at all include instances where failure has occurred due to lack of maintenance by the owner when he was aware of what should have been done. At other times, failure may be the consequence of natural phenomena which the designers and contractors could not reasonably have been expected to foresee.

For example, there are varying degrees of storm conditions which should be taken into account when buildings are designed. However, if a storm of exceptional and unprecedented severity were to occur, resulting in damage to several buildings in its path, such damage would be less likely to be regarded as the consequence of poor design or construction.

To take another example, consider a material in common use which had been tested by responsible agencies and was considered by the industry as a whole to be acceptable. Failures resulting from the widespread failure of the material as a whole (rather than a faulty batch) might well give no grounds for action against designer, contractor, manufacturer or anyone else. In other words, if the 'state of the art' at the time of design and build was universally recognised, and only subsequently did widespread failure followed by research indicate fundamental misconception, action would be unlikely to succeed. For example, for several years high alumina cement was believed by the building industry to cause no deterioration of the concrete it helped to make. The failures that in some circumstances occurred as a result of the use of this material during the period it was accepted by the entire industry as satisfactory could not be blamed on designers relying on universally accepted 'state of the art' knowledge.

'Economic' considerations influencing decisions

There are, however, other circumstances where however good the technical and legal cases there would be no point in proceeding with legal action or arbitration. Take the situation where it is established that failure is due to faulty workmanship but the contractor has gone out of business due to insolvency. The only claim against the architect would be for the cost of correcting such bad workmanship as he could have been reasonably expected to pick up during inspections of the work in progress. If the architect was a one-man practice with no insurance against possible claims, it might not be worthwhile attempting a claim.

This example leads to the general principle that when considering making a claim, the likely cost of proceeding with it should be equated to the sum one could expect to receive if the claim is successful. The result should then be considered in the light of the hidden costs in terms of money, management time, effort and stress that are bound to be incurred along the way, as well as tangible irrecoverable costs. Depending on the size of the claim, these could be of very considerable significance. It should also be noted that a successful litigant is unlikely to recover more than 80% of his costs, and the figure may be as low perhaps as 60% on occasion and, exceptionally, even less.

Legal influences

There are many occasions when in technical and financial terms action is justified, and in these cases the next step is to assess the likely chances of success in legal terms, for both arbitration and litigation. Assuming that the proposed defendants exist and are financially able to meet the claim and costs, the decision regarding action becomes influenced almost entirely by legal issues. Although this is not a legal book, some of the major points for consideration must be highlighted as being relevant to the making of decisions which may have far-reaching consequences.

Legal grounds for action provided damage has been suffered

An action against a second party or several parties (the defendants) may be on the grounds of an alleged breach of contract or it may be in tort (a wrong). As examples of each consider first a fare-paying bus passenger injured by the negligent conduct of the driver. By virtue of tendering the fare, being given a ticket and accepted as a passenger, there is a contract with the bus operator. If, however, the person is knocked down by a bus while crossing the road he or she would have no contract, and any claim for damages against the operator would have to be in tort.

In the case of building disputes, which of the two grounds is appropriate depends on the circumstances of the individual case and is very much a matter for the lawyers. Both are subject to limitations in terms of time. Depending on the overall circumstances, including among other things whether contractual breach or tort is alleged, and whether a contract is under hand or seal, liability may extend to a maximum of 15 years. The period is also related to the date of the commencement of damage. The situation for claims in tort is dealt with in the Latent Damage Act of 1986. However, the whole question of limitation is highly complex, and legal advice is essential. The advice should be sought as soon as possible after it is established that responsibilities for the damage may lie with one or more members of the design and build team. A thing sometimes difficult to pinpoint is the date that physical damage occurred. For example, if a cavity wall collapses due to the corrosion of wall ties, does damage occur on the day that the wall collapses after a heavy storm, or when the ties have become ineffective as a means of making the two skins act as one wall? If the latter, at what point had enough ties suffered enough corrosion to allow the failure of the wall during the next severe storm?

The position of subsequent owners

Recent legal decisions mean that if no contract exists between plaintiff and defendant it is much more difficult for the plaintiff to recover losses. Generally he will only be able to do so to the extent that he can demonstrate that the building is, or is likely to become, unsafe. Like many of the statements in this book relating to the law, this is an extreme simplification of a matter which is in fact highly complicated and far from clear. It is the subject of much discussion, if not argument, in legal circles. Therefore, where the owner of a building is not the original owner, there is likely to be no contract between the owner and any members of the original design and build teams. Subsequent owners find themselves in difficulty therefore as litigants unless they can satisfy a court of danger or potential danger in their building, responsibility for which can be pinned on the original designers and/or contractors and sub-contractors.

There is now a tendency for purchasers of existing buildings to try and obtain a warranty from the vendor covering the possibility of latent defects surfacing in the future. In consequence, the developers of buildings are attempting to obtain collateral warranties from the designers and builders which can be passed on to a future buyer to enable them to sue if necessary. The warranty would preferably therefore contain provision for the contractual responsibilities of the designers and builders to be assigned to such future owner.

During the initial appointment negotiations between the original client and the consultants, the former will be doing his best to keep the warranties as exacting as possible, and the latter will be doing the opposite. Subsequently, when tender documents are prepared, efforts will certainly be made to make the contractors responsible for as much as possible, with the contractors again doing their best to keep future responsibilities to the minimum. No doubt market pressures will play their part in the final outcome as far as the basic acceptance and the terms of warranties are concerned.

At the present time, unless suitable warranties exist in any particular instance, a subsequent owner will have little option but to sue in tort, with a considerable reduction in the chances of success compared with a claim in contract based on the terms of the warranty.

A building owner who is not the original initiator of the project should, in the same way as everyone else, obtain legal and technical advice as early as possible after the realisation that he has problems. He should liaise with his experts at all stages and give them as much information as he can as early as he can, whether or not it seems relevant at the time.

Information must be evaluated by advisers

It should be left to the experts to decide what is and what is not relevant. Also, the building owner must resist the temptation to suppress information because he thinks it may damage his own case. Possibly because he has perpetrated stupidities and (a very human failing) does not want to appear stupid to the world, or for other reasons, be they sinister or not. In the long run the advisers, both legal and technical, can only do their best for a client if he gives them all the information in his possession that has any bearing whatsoever on the matter in hand. This, of course, applies to all involved, be they plaintiff, defendant or third party.

Chances of success

There are at any time a number of matters which may affect the outcome of an action, quite apart that is from the technical merits of a case and the financial standing of the potential defendants.

Among other things, the law tends to change as 'case law', the result of court judgments, is referred to and taken into account in future trials. This is quite apart from actual legislation.

As investigation and consultation progress, the experts, both legal and technical, will build up a picture of events that will enable them ultimately to recommend to their client the advisability or otherwise of some form of action to recover the cost of damage. This recommendation will probably be based on an opinion by counsel who will, given the full picture from all the other experts, be able to make a judgment on the likelihood of success.

What type of action?

Should action be by negotiation, arbitration or litigation? The merits of each, and the situation should negotiation be tried and fail, are discussed in the following chapter. Before dealing with them, note that as far as court action is concerned, the situation does vary depending on the size of claim. At the present time, small claims (small that is in the eyes of the courts) of less than about £5000 may be dealt with at County Court level rather than in the High Courts, with usually a considerable reduction both in terms of costs and delay, but with the judge having less expertise than an official referee. It is also possible for claims higher than £5000 to be referred to a County Court if considered appropriate. It is, however, very seldom indeed that such small claims will be worth pursuing or fighting, but it does occasionally happen.

Chapter 8

Negotiation, arbitration or litigation?

Settling a dispute without recourse to either arbitration or litigation will save all concerned much time, stress and money. With some disputes, however, there is often one party who is convinced that his case is so good that he will save money by refusing to negotiate. Sometimes, though, such a person will accept that whatever the rights or wrongs of the matter, the amount of money involved, when related to other relevant factors, makes settlement by negotiation the economic and common sense answer. (On the other hand, professional indemnity insurers acting on behalf of the professionals of the design team frequently seem reluctant to settle under any circumstances until they are within sight of the court door.)

Encouragements to settle may include the avoidance of disruption, time, stress, cost of damage related to irrecoverable costs and the maintenance of good will. The principal reason is normally, however, because it seems a sensible course financially.

Very occasionally a case may be so strong and so obvious that one party really has no option but to accept responsibility. On the other hand, a person may feel so strongly about an alleged wrong, as either complainant or alleged villain, that they will refuse to concede or negotiate whatever the dictates of common sense. Sometimes the amount of a claim may be less than the irrecoverable costs incurred by the parties!

More damage equals less chance of agreement

Generally, the greater the percentage value of the alleged damage related to the whole, the less likelihood of agreement without argument. For example, a request to correct damage due to alleged faulty workmanship requiring expenditure of £5000 immediately following the completion of a multi-billion pound contract will frequently be accepted after little more than token resistance. This is particularly so if there is a possibility of major future work in prospect.

The same amount at stake on a contract of £50 000 would be likely to produce a very different reaction.

What is the evidence?

No one is likely to admit liability without adequate demonstration of the strength of evidence. This will mostly require an expert report. In the case of faulty workmanship, a report from the original designer may be accepted, but frequently in the case of workmanship, and always if the standard of design is in question, an independent report will be necessary as far as the owner is concerned.

Legal advice is essential

Even if a negotiated settlement is contemplated from the outset legal advice will be necessary. It would be foolish for any party to attempt to progress any sort of action, be it any of the three which form the subject of this chapter, without legal and technical advice as to the most sensible course of action to be followed. Apart from any other consideration, there is the 'clock-stopping' advantage, described earlier, of issuing a writ, even if not serving it, which could later prove to be vital from the time limitation point of view.

In the early stages, action by a building owner after the incidence of damage to his property tends to follow the same pattern, whatever the subsequent decisions regarding the progression of events. The decision regarding the choice of course to be adopted may, and probably will, involve solicitor, barrister and technical expert. The choice should, in the majority of cases, only be made after a full assessment of the damage and its causes has enabled the technical expert to be reasonably certain on the question of who is (or are) in his opinion responsible, and the lawyers have been able to digest this conclusion and relate it to the other relevant circumstances of the case.

Settlement negotiations

How discussions with a view to settlement by negotiation are instituted and subsequently handled is dependent on the overall situation. Even if settlement is unlikely in the early stages it will (and should) usually be attempted prior to the delivery of a 'letter before action' and the service of a writ. Achievement of settlement at this early stage is likely to depend on the common sense economics of the situation and/or the other factors outlined at the beginning of this chapter.

It is not uncommon for intended defendants to go some way forward with defences or counterclaims, even when they know that their case is not particularly strong, if only to test the resolution of the claimant and the skill of his advisers. Where designers are concerned, professional indemnity insurance is a likely factor. Unsurprisingly, the insurers of a potential defendant will want a considerable input into the manner in which the insured's affairs are conducted, almost certainly to the choice, or at least approval, of legal advisers and technical experts, and the conduct of the defence of the professional defendants whom they have insured.

So, even if settlement by negotiation is hoped for by all parties, it is likely that matters will proceed some way along the paths of either arbitration or litigation before settlement is reached. (Most disputes are resolved before or during a hearing.)

Before a discussion of either there are two basic situations to be taken into account. The first, where the symptoms of failure occur during the defects liability period following practical completion of a contract and the second, where symptoms only manifest themselves after the end of that period. In the first case the contract will normally still be within the control of the original designer. It is very unusual indeed for a designer, faced with symptoms of failure, to accept design faults by him as the cause. They may be ascribed to poor workmanship, faulty materials, lack of maintenance or, as has been known, to 'the building settling down'. It is the second situation that is the norm as far as this book is concerned.

If there is an admission of design error, it will very probably be one perpetrated by 'one of the other consultants'. To the contractor, however, the problem is almost always the consequence of poor design or maintenance! To the owner, anything but poor maintenance or misuse of the building. Unless, of course, it is let on a fully repairing lease.

If early attempts at settlement fail, there are the two remaining options to consider.

Arbitration

In the past there has actually been some litigation stemming from the lack of any statutory definition of arbitrator, and, indeed, arbitration. Such litigation has been brought about by the fact that the law seems to suggest that arbitrators are immune from actions brought on the grounds that they have been negligent or lacking the necessary expertise to carry out their function properly. The most that can happen is that, if the arbitrator errs on a point of law, his judgment may be set aside by the courts, and the parties could be almost back where they started in terms of an award or judgment, but way beyond in terms of the expenditure of time and money. However, even obtaining permission from the court to challenge the award of an arbitrator may prove difficult.

As far as building is concerned, there is invariably a provision in a contract between employer and contractor for contractual disputes to be the subject of arbitration. The arbitrator will be appointed by agreement between the parties or, if they cannot agree, by a person nominated in the contract to do so. There may or may not be arbitration clauses in agreements between the employer (client) and his professional consultants, but such provisions are on the increase.

Usually, a chosen arbitrator will have far more technical knowledge of the subject in hand than any official referee (an official referee is a judge who specialises in the hearing of technical cases related to building and civil engineering) or other judge can, apart from exceptional circumstances, be expected to have.

Conversely, no arbitrator can reasonably be expected to have the legal knowledge possessed by a judge. Whatever the experience of each with regard to the expertise of the other, there may be some difficulties. The judge perhaps has more opportunity to be advised on technical matters during the actual progress of a trial than the arbitrator of legal matters during the actual hearing of an arbitration. Nevertheless, the latter would normally have ample opportunity to obtain any necessary legal advice before committing himself to a written award.

It is not always possible to make a choice between arbitration and litigation, as one or other may be dictated by circumstances. For the occasions when it is, the pros and cons are as follows.

One of the oft quoted advantages of arbitration is that in addition to the technical knowledge of the arbitrator, proceedings are much less formal. In fact, if lawyers are involved the proceedings are very similar to court. Others are that there is a much greater degree of privacy, and a shorter time lag between decision and hearing. The time and place may be agreed between the parties and the arbitrator rather than being imposed by the court, with a saving in cost (doubtful, as the hall, arbitrator and shorthand writer all have to be paid for) and also the parties may be represented by whom they choose. In the case of litigation the situation in early 1990 is that representation may only be by barristers. Whether this will change in the near future is a matter of conjecture, it has of course been mooted. Also, legal aid is not currently available for arbitration. One of the disadvantages of arbitration is the inability to join a third party. It is possible, but highly improbable, that if all parties agree, two or more separate arbitrations may be run concurrently by the same arbitrator to give a situation similar to the involvement of a third (or fourth) party.

Another disadvantage is that an arbitrator has no power to order either party to keep to a timetable with regard to production of reports and other matters. This may result in a reluctant participant being able to drag out the proceedings for a considerable time. The arbitrator may set down a date for the hearing, but even then may find it difficult to enforce if the reluctant party produces even moderately convincing reasons for delaying the production of essential documents. Generally, in fact, the arbitrator has less power than a court to put pressure on parties to avoid delays. (The reader is referred to *Arbitration principles and practice* by John Parris (see bibliography), 1.13 of Chapter 1 sets out fully the advantages and disadvantages of arbitration as opposed to litigation.)

Arbitration is binding

An arbitration is binding on both parties and there can be no appeal except in the case of a legal error by the arbitrator. If a contract makes no mention of possible future failure, in such an eventuality it should still be possible to appoint, by mutual agreement, an arbitrator, should the dispute be limited to two parties. If the building owner believes that more than one party has contributed to his problems, it is, in theory, possible to adopt the course described in the previous section. In the great majority of cases, other than the very simplest, both situations seem highly improbable as they would require very considerable goodwill and co-operation between several opposing parties.

It is unusual for any sort of agreement to be reached at an early stage that would limit responsibility for a complaint to one party, or even two. This would tend to militate against arbitration being acceptable or a single arbitration even feasible in all but a tiny minority of small value claims where only two parties were involved anyway.

Who can act as arbitrator?

There are no essential formal qualifications, and in theory the main requirement is that the arbitrator should be a person acceptable to both parties to a dispute. In practice, as far as building is concerned, it is common (and sensible) for the person agreed and appointed to be a Fellow of the Chartered Institute of Arbitrators. The rules of the Institute do not allow for Associates to conduct arbitrations. To act as an arbitrator an Associate must achieve Fellowship by succeeding in the examinations set by the Institute for this purpose. Most arbitrators are also members of a profession relevant to the nature of the dispute. With regard to building matters, they would usually be architects, engineers, surveyors, builders or lawyers. The first four selected for their technical knowledge of building, the last if a dispute is predominantly of legal rather than technical issues. There are a few practitioners with both technical and legal qualifications, but they are comparatively rare. Most of them concentrate on either the technical or legal side, but obviously the second discipline is likely to be of considerable value. It is also possible to nominate an official referee as arbitrator.

Independence of the arbitrator

When disputes arise during the course of a contract and before the end of the defects liability period there may well be situations where the supervising officer, in the course of his proper administration of the contract between the employer and the contractor, will at times be virtually acting as a mediator between them, although this is not his proper role. In such circumstances it is usually accepted that the officer will be impartial despite being employed to look after the interests of his client, who is the employer in contractual terms. If a dispute reaches the proportions of needing the appointment of an independent arbitrator it is sensible, even if not a legal requirement, for him to be just that – independent. He should not be biased against either party and neither should he have any involvement that could possibly require him as a witness. In other words, he should be a person entirely unconnected with the project, and not even someone who has given advice on design or workmanship matters before or during the progress of the work.

Appointment of arbitrator

The arbitrator must be appointed in accordance with the requirements of the contract, and the appointment formal and properly documented. It is strongly advisable for the whole process to be overseen by the owner's legal advisers to try and obviate any legal error by the arbitrator that could result in considerable delay and further expense should the arbitrator's award be set aside.

Future contracts, employer/consultants, employer/contractors

Currently, there are indications that contracts between employer and professional consultants are likely to become far more detailed, comprehensive and specific in defining the duties and responsibilities of one party to another, and will include clauses relating to the initiation of arbitration in the event of a dispute, in a similar manner to employer/contractor contracts.

Litigation

Appeals

In the event of a settlement or at the end of an arbitration even the loser may feel a sense of relief that it is all over and done with, despite a probable requirement to part with substantial sums of money in addition to what has been spent on his own costs. At the end of a trial there may well be an appeal to be heard in due course by the Court of Appeal, and that followed by an appeal to the House of Lords. Fortunately, appeals occur only in a minority of cases, and, as mentioned, the initial trial frequently does not run the full term or even start. More often than not, a negotiated settlement will be arrived at shortly before a hearing is due to start or shortly after it has started.

Delays

Perhaps the principal disadvantage of litigation compared with arbitration is that at the present time litigation tends to be much slower in coming to trial due to the large number of cases down for hearing related to the number of official referees available to hear them. It is understood that efforts are currently being made to increase the number of official referees available.

Currently, it is common for the official referee's fixture lists (cases awaiting hearing, with anticipated dates) to be double, treble or even quadruple booked on the premise that the majority of cases settle before they reach court. However, if fixtures at the top of the list do run, those below will have to be put back to new dates, albeit with a higher priority than before.

Litigation permits more than two parties in one action

One of the advantages of litigation as opposed to such disputes as can be settled by arbitration is that all the parties likely to be involved can be brought without difficulty in to the one action. Unlike arbitration, an action can be brought against any number of defendants, and the defendants can join (bring into the action) third parties as they feel necessary. For example, if a building owner considers that both designer and contractor are partly responsible for his problems he can claim against both in the same action. Also, if one designer, the architect perhaps, feels that he was wrongly advised by, say, an engineer, he can join him as a third party. In a similar manner, the contractor may join a sub-contractor, who in his opinion is responsible for poor workmanship for which he, the main contractor, is being sued by the building owner. Such a facility is, as earlier demonstrated, extremely difficult in the case of arbitration.

This has in some cases led to a 'sue everyone in sight' philosophy, but it does in the main mean all can be settled at one fell swoop. There are reasons for 'sue one, sue all' which will be discussed later. Another advantage of litigation is that the judge does not command a fee or the court a rental. Against that, cases large enough to be heard by an official referee rather than a judge in a County Court are mostly heard in London, with some in centres such as Liverpool, Manchester, Bristol and similar large and strategically located conurbations. This could involve some litigants in the considerable costs of travelling, travelling time, and hotel accommodation for often quite large numbers of people. The cost of transporting large numbers of documents can also be considerable. Much will depend on the overall size of a claim. In the same way that it can take as long for an architect to write a specification for one £200.00 window as for a hundred similar windows, so the amount of documentation in a trial is not always directly proportional to the amount of the claim. This applies just as much to arbitration as it does to litigation.

Comparative costs of arbitration and litigation

With regard to costs, savings with arbitration rather than litigation are only likely in contractual disputes where claims are quite small. Once claims are substantial, costs will include not only the arbitrator's fees, etc., but all the costs associated with litigation such as solicitors', barristers' and technical experts' fees, photocopying, photography and stationery (all of which can mount up to significant sums), and all the other expenses involved with litigation. In some situations, however, there may be quite large savings connected with travelling, as well as earlier resolution of the dispute than would be the case with litigation, with the need to get to the top of a waiting list. This latter situation will depend of course on the availability of the arbitrator and the degree of co-operation between the parties.

Arbitration or litigation? Summing up

All in all, arbitration at the present time may be suitable for contractual disputes which in some instances may involve building failure. This is true at least where the sums involved are comparatively small and the dispute can be clearly confined to two opposing parties. Where the claim is substantial, the cause is failure leading to damage and loss and the responsibilities are not clear-cut and lie with more than one party, arbitration is unlikely to be a satisfactory method of resolving the problems of the dispute. In such instances litigation, but with settlement in mind, is often the logical way to proceed.

Chapter 9

The position of the defendants

So far, consideration of events has dealt largely with those generated by the building owner and his 'failure' team. It is timely now to think about the reaction of the other parties, those alleged by the owner, as potential or actual plaintiff, to have been responsible for the problems. Also, the defendant may well believe (or purport to believe!) that he has done nothing wrong and everything right. Should this be so, he will either not accept that there are faults, or he will try and move the responsibility for them to others. This may be to any one or more of the design and build team, to checking agencies (Local Authority Building Control for example) or perhaps to the plaintiff himself. In the latter instance, there may or may not be, depending on circumstances, a counterclaim.

Who will the defendants be?

Mention has been made of the tendency in some cases for the plaintiff's lawyers to suggest suing anybody and everybody who may possibly be concerned with the incidence of damage. To initiate an action there must be some tenable evidence against a potential defendant, but even if it is only indicative of a very small degree of responsibility there may be sound reasons for an initial writ and ongoing pursuance of a claim. One reason for involving parties with relatively small responsibility is that quite often it is in the commercial interests of a potential defendant to settle a claim long before trial. This may seem in many cases to be hard when the original claim and the final settlement sum are in excess of what is genuinely thought to be the limit of liability, but as long as it is a sensible economic answer it will probably continue to happen. However, in the first instance there must have been some real basis for a claim or it is unlikely that any organisation with the resources to be worth suing would entertain any thoughts of capitulation without a fight.

Usually it will not be until all technical investigations have been completed and all the relevant documents have been made available to, and studied by, the plaintiff's team that responsibilities can be determined. Sometimes limitation problems may make it necessary to establish likely responsibilities, at least, at an early stage.

Liability for damages against other parties

If two defendants are found liable for damage, and one is without financial resources adequate to meet the damages imposed, the other party will be liable for the full amount of the damages, irrespective of the proportion of blame which he bears. For example, if there is a total judgment of £500 000.00 against defendants A and B, then those two defendants will each be liable in full to the plaintiff for that amount. Thus, if B is considered 10% to blame and A 90%, A will be given the right to recover £50 000 from B (on payment by him of the judgment sum to the plaintiff) and B will be given a similar right to recover £450 000. If A and B are both capable of meeting their obligations this is largely academic. If this is not the case, the situation changes radically. Thus, if A were to have insufficient resources to make any payment B would be obliged to pay the plaintiff £500 000 although only 10% to blame. His right to recover £450 000 from A would still exist but would in practice be worthless. In this way, when more one than one defendant is sued, it is normally the other defendants rather than the plaintiff who take the risk of the insolvency of one of their number. The court may apportion the total sum between the two defendants according to their degree of culpability.

Owner's invitation to alleged defaulters to pay for damage

To recapitulate briefly the likely situation when a building owner is contemplating legal action or arbitration. He will have suffered loss in legal terms at least and, hopefully, will have obtained a report from a technical expert giving opinions on the cause(s) of damage, responsibilities and method and approximate cost of remedial work. He will also have obtained legal advice regarding the next steps. Whether arbitration or litigation is in mind, it is highly probable that those thought to be responsible for the problems will be so advised by the owner's solicitor, and invited to pay for the cost of the remedial work and any other costs incurred by the owner as a direct and unavoidable consequence of the damage.

In most instances this will not come as a bolt from the blue. It is likely that there will have been some expressions of dissatisfaction that have reached the ears of the potential defendants. It is probable that they have already been invited to try and diagnose the cause of the problems, and perhaps made unsuccessful attempts to put them right, or alternatively, have claimed that they are not their responsibility. Maybe the owner's expert, only, it is stressed, with the blessing of the lawyers, has already contacted some, if not all, members of the professional design team and/or contractor to try and obtain drawings, specifications, etc., to assist the diagnostic process.

Involvement of insurers

It is likely, but not certain, that once there is a clear indication of trouble ahead insurers will be notified. Most consultants carry professional indemnity insurance. Professional indemnity and similar insurers invariably insist on an 'early warning' of a possible claim, without which the insurer may well disclaim any obligation to meet any damages and costs imposed on their clients. Such warning or indication is usually required as soon as the potential defendant has any reason whatsoever to believe that there are circumstances which may give rise to a claim. If a client has any reason to believe that his designers may be responsible to any degree for the defects in his building he should make certain that they notify their insurers as soon as possible! This is a matter to be weighed against any possible advantages of the owner not giving the designer early notice of a possible legal claim. Certainly, at times designers show a reluctance to inform their professional indemnity insurers of a possible claim, only exceeded by the eagerness of the latter to repudiate it if there are any grounds for doing so.

Intended defendants and their need for advisers

Another point is the timing and complexity of the owner's expert's investigations. It may well have been necessary to alert the designers and contractors at a very early stage of the likelihood of a dispute to enable the proper diagnosis and possibly execution of emergency remedial work. Also, it will be recalled that the owner has a duty to mitigate his loss, and this could necessitate very early advice to potential opponents because of the need to carry out remedial work which would destroy evidence of the causes, real or alleged, of failure.

Anyway, at some stage in the work of the owner's expert the

potential litigants (henceforth, reference to litigation and those involved includes arbitration) will need legal advice and the services of an expert or experts of similar disciplines to those employed by the owner. Precisely when this stage is reached will depend on the individual circumstances, but as a general rule, the earlier appropriate professional advice is sought and followed, the better. It cannot be overstressed that trying to 'go it alone' is, with very few exceptions, a recipe for disaster. This applies even if the intention is to call for the correct help after initial attempts to sort the problems unaided, as irretrievable damage may be done.

Defendant's choice of advisers

How then to choose the advisers? The way forward for the defendant(s) was dealt with as far as some aspects are concerned in Chapter 3. The part recapitulation, part new ground that follows is intended to put all into context.

Most firms, of whatever discipline, be they professional consultants, contractors or suppliers, almost invariably employ a solicitor to handle their legal affairs, and that solicitor is likely to be the subject of the first approach to obtain specialist help. There are times, however, when advice may be sought initially from a fellow practitioner or contractor known to have experience of building failure, be it as ex plaintiff, ex-defendant or as an acknowledged expert. Alternatively, in the case of consultants, it may be that the insurer, having been advised of looming trouble may suggest, and often insist, on the use of particular legal and technical advisers known by them to have specialist knowledge of the problems at issue. At the very least, they will usually require to vet the advisers before approving their appointment.

Whichever route is taken, it is possible that whoever is approached first may be in a position to recommend firms with appropriate expertise and experience. As far as solicitors are concerned, the protagonist's own solicitor may have the requisite experience or be in a position to find out who has.

Registers of specialists

As for the plaintiff, use may be made of the registers kept by the various professional bodies to find suitable legal and technical advisers. Alternatively, recommendation by a fellow practitioner or contractor may be possible, while once one adviser has been found, he will usually be able to recommend suitable members of the other disciplines involved. All in all, the process of choosing and appointing the team is very similar to that used by the building owner, with one fundamental difference. The building owner may start by not realising that his problems are going to be anything other than technical (with attendant financial implications), whereas at the stage now reached the choice of adviser is made with the knowledge that litigation is in the offing.

It is still vital to spell out exactly what the problems are, the details of the project and the parties involved. It has happened in the past that extreme coyness on the part of a potential defendant has resulted in time and money being wasted before an expert who has been approached has been able to establish that the building in question is one for which he has already been commissioned by another party to the same dispute!

However, for experts or indeed solicitors to act against a party in one case, and for them in another later on, is certainly by no means a rarity. This by the way should not be taken to imply that the whole of the building industry is in a continuous litigious state. It does however happen over a period of time that someone may be involved in further litigation and have been sufficiently impressed by an adviser on the other side in the earlier case to prefer to have him as an ally if the need should, unhappily, arise.

Adequate briefing essential

Once an expert has been selected, approached, and he has agreed to accept the commission, he must be carefully and comprehensively briefed. It is not surprising that any party to a dispute will attempt to present his past actions in the best possible light to anyone representing him, including, let it be said, himself. Barristers, solicitors and technical experts, apart from being almost invariably extremely conscientious, are all aware that it is in the interests of themselves, as well as their clients, to do the best they can to give the correct advice. This extends to telling someone that he has not got a good case should that actually be the situation. It is madness to proceed with litigation if one is bound to lose, and objective and useful advice can only be given if the adviser knows all the facts, the bad as well as the good.

Even if knowing the bad does not result in advice to admit defeat, it is most important that lawyers and experts are not presented at a late stage with some damaging revelation that could have been dealt with comparatively simply if known about at an early stage. No one likes to admit that they have been incompetent, foolish, lazy, negligent or even less than honest. Nevertheless, advisers cannot without difficulty minimize the effects of such conduct if they do not know about it until late in the day, sometimes as late as cross-examination at trial. The need to avoid being economical with the truth applies, of course, as much to plaintiffs as to defendants!

However, there is more to briefing an expert than showing the warts. It is important for him to be given the fullest possible information available to those commissioning him. As much as possible of what is known about the allegations made, the evidence on which they are based and all available relevant documentation. What too is the expert being asked to do? He should try and arrange to be present at site investigations not yet carried out and should be aware of the need, if any, for the presence of supporting experts and recommend their employment and liaise with them at the right time. That is, as far as possible, before things happen, not after. They should know who is acting for other parties and what are the lines of communication between all parties. These are often via solicitors, but there is frequently, if agreed by the lawyers, a degree of communication directly between experts on some matters.

Whatever the arrangements the expert must know them. Most experts, when appointed, will produce a list of questions, the answers to which will form a large part of their brief. One question that invariably crops up of course is that of fees. These will usually be on a similar arrangement to the plaintiff's expert, i.e., on a time basis. One difference though is that a defendant's expert is even less able than the plaintiff's to control the time needed. He will mostly need to attend investigations and inspections arranged by the other side, and to deal with documents not of his own making, apart, of course, from his own report.

Who appoints the expert?

Once the defendant's expert has been selected and briefed the question of who actually appoints him will need to be settled. It is possible for him to be appointed and paid directly by the defendant, by the solicitors on behalf of the defendant or by the defendant's insurers. It is necessary for the expert to know definitively for whom he is acting, from whom he should take his instructions and by whom he is to be paid! Whether or not the solicitor, it is important that work of substance is only undertaken (or not undertaken as the case may be) after consultation and with the agreement of the expert's client and usually the other main members of the client's team.

Technical expert appointed before lawyers

There are situations where an expert may be appointed by a potential defendant prior to the involvement of lawyers. It may be that someone such as a building owner is making a claim against a contractor, or an architect instructing a contractor, sub-contractor or supplier to correct allegedly faulty workmanship or replace allegedly faulty materials. In such cases an independent expert may sometimes demonstrate successfully to the claimant that the claim is ill-founded and, as far as the supposed defaulter is concerned, that is the end of the matter. The manner in which the claimant is convinced may well lead to the initiation of action against someone else, but that is another story.

Timing of expert's appointment

There are various stages at which a defendant's expert may be appointed. As a rule, the initial information relating to damage suffered by the plaintiff will be contained in a writ, very probably, but not necessarily, supported by a rather more detailed statement of claim.

It is unlikely that initially a comprehensive report by the plaintiff's expert will have been disclosed to the defence team. If the plaintiff has not fully established the causes of the failures suffered it may be that all experts will be present when detailed site inspections and investigations are carried out. This is preferable in every way to the defence attempting to build up an accurate picture from records, however good, made previously by someone else. It is unfair to expect defence experts to do their job properly without an opportunity to see for themselves at first hand what is claimed to be wrong and the alleged causes.

Nevertheless, for the reasons mentioned earlier it is not always possible for the defence experts to be appointed early enough to see at first hand all the failures, symptoms and suggested causes. Where this situation pertains it is essential as far as the plaintiff's expert is concerned that the reasons are sound and that evidence is clear, adequate in extent, and everything likely to be significant in the eyes of all involved is properly recorded and as far as possible verifiable as a true record. Certainly the defendant's expert must bear this in mind and take all reasonable steps to see as much as possible for himself. However, it is repeated that there is no legal requirement for the plaintiff to advise potential defendants of site investigations and/or give them an opportunity to be present, although it is normally in the interests of the plaintiff to do so. The reader is referred also to the paragraphs in Chapter 4 concerning site investigations.

Involvement of third parties

At a comparatively early stage after the initiation of action against a member of the design and build team, the defendant may consider that his direct responsibility for design deficiencies on the one hand, workmanship on the other, should really rest, wholly or partly, with another member of the team.

Take for example the situation where, on a small project, the employment of a structural engineer by the client is not justified but the architect takes advice from an engineer for the design, say, of one beam. Failure of the beam is alleged to be the consequence of poor design. The building owner may sue his architect, and the architect then join as a third party the engineer who in fact did the design work, but who, as far as the owner is concerned, does not exist.

A similar situation could well apply with respect to faulty workmanship or design carried out by a domestic sub-contractor. Such a sub-contractor, responsible in some degree for the problems, normally has no contractual relationship with the owner/employer. Again, the sub-contractor may well be joined as a third party by the main contractor, and the third party can join a fourth! It may transpire that the sub-contractor is wholly responsible, or partly perhaps, if it can be established that his work should have been checked by one or more other members of the design and build team.

Liaison between opposing experts

Whether first, second, or other defendant (remember the plaintiff may well consider more than one party to be directly responsible in some degree), or third party, each of those sued by the plaintiff and joined by a defendant will need to appoint their own expert. As outlined, there may be, in complex cases, several experts for each party, though normally only one of each discipline per party.

When the principal expert for each defendant has been appointed, and those acting for third parties (if, in the early stages the third parties have already been joined), the plaintiff's expert will normally inform them of the procedures to be followed. This may be done via the solicitors acting for each party, but once initial contact has been established the arrangements for site inspections and investigations will often be made directly between the experts, but within the constraints of agreed guidelines.

For example, the plaintiff's expert will usually try to arrange investigations on site at dates and times suitable for all experts whose attendance is requested by their own side. Sometimes it may not be possible to fall in with everyone's wishes, and provided a reasonable attempt has been made and reasonable notice given, it may be necessary to go ahead with visits that do not suit everyone. Should that happen, it is important for all sides that the circumstances are placed on record to avoid vexatious claims and counterclaims by the various experts relating to the opportunity, or lack of it, to see things for themselves. Remember that other people will have to be considered, for example, occupants of the building, remedial work contractors and, in some cases, members of the public.

When site investigations are carried out they are invariably under the control of the expert to the plaintiff, with the others as observers. However, experts will accommodate the reasonable requests of the opposition's representatives. This will extend, for example, to taking moisture and humidity readings, taking photographs, checking dimensions, removal of a reasonable number of samples and similar activities.

The defendant's expert working towards his report

Once the defendant's expert has had an opportunity to study the statement of claim in detail and also to witness site investigations he will begin to build for himself a picture of the symptoms of failure and of the actual damage. These, combined with a study of available documents, should help him to form quite quickly an indication of causes and responsibilities. There is however no guarantee that he will be able to do so and, whatever happens, any first thoughts which are not fully substantiated by hard evidence should be suitably qualified when expressed to his client and his client's lawyers. Obviously, the sooner he can give an accurate indication of the chances of a successful technical defence to the claims the better.

Expert's early advice to the defence team

Frequently, both defendant and his lawyers will be tempted, understandably enough, to pressurise the expert into giving an opinion. The prudent expert who is not certain will try to refuse. It is an unfortunate fact of life that when qualified opinions are given, as time goes by the opinions are remembered and the expressed qualifications forgotten. The same thing happens with approximate estimates! The expert therefore should beware, and, better still, defendant and lawyer resist in the first place the temptation to do other than ask for an opinion as soon as possible, with updates on progress at suitable intervals. It is of course incumbent upon the expert to appreciate the anxieties of his client and come up with the answers just as soon as he can do so with an assurance that will stand up in court. Having said that, early opinions will continue to be sought, and experts will continue to give them. Hopefully, for the sake of everyone, with both opinion and any qualifications to it in writing.

The expert's preliminary report

As the owner's expert has done before, the defendant's expert will, as early as practicable, produce a preliminary report that will as far as possible enable the defendant, in consultation with his advisers, to decide on his next move. As the choices are likely to be between negotiate towards settlement or fight, it is not difficult to understand why the expert has been put under pressure to produce the technical answers. These of course have to be considered in conjunction with the legal aspects of the situation. The expert will be doing his best to give his opinion as early as possible. In the real world the too early, unsubstantiated opinion may be inevitable, but it is repeated without apology that the vital need is to explain to the full what is unsubstantiated and why. Also, what else needs to be done before it can be definitive. Only if the defendant and the lawyers know what is involved can they make a sensible decision regarding the next move. It may well be that the most sensible action is to proceed with investigations that can make the outcome, if not clear-cut, at least reasonably predictable.

Very often, the preliminary report may be a phone call, confirmed in writing, explaining the need to do more work, or have work done by specialists, and seeking the authority for it to be done. On the other hand, it could be at a meeting between the defendant, expert and lawyers. If the original brief has been sufficiently wide the preliminary report could be a comprehensive statement of the overall position, needing only an update and perhaps amendment as documents come to light, perhaps after disclosure by other parties in the proceedings. It is better to deal with the defendant's and third party's further reports in the ensuing chapter, apart from the reminder that, as with the plaintiff, the status of a report can best be positively identified by date, which is completely unambiguous when referring to it in a letter or during a phone call.

Chapter 10

Pre-trial events

Mention has been made of some of the events and documents that respectively occur or come into being between a symptom of failure being seen and the start of a trial. Most of these need discussion in greater detail, and others are still to be introduced. Not all of them are always needed.

Issue and service of writs

It will be recalled that a writ may be issued as a precautionary measure to 'stop the clock' where a proposed action could become time barred. Such a writ, in the light of further detailed examination of the problems, might not ultimately be served on a particular party. Assuming that early on writs had been issued against a number of parties who might later be considered to bear some responsibility. Such writs may eventually be served on none, one, several or all of the original possible defaulters. It could even happen that a completely different person could emerge as a principal or subsidiary defendant. Remember too that a writ, once issued, ceases to be valid unless served within 12 months of issue, and that service of a writ is accompanied or followed by a 'statement of claim' giving more detail of the allegations. It is sometimes possible to apply for permission to renew a writ. Even if permission is granted by the court, difficulties may be caused with regard to the Limitation Act.

Service of a writ has to be acknowledged within 14 days if accompanied by a statement of claim, or, if not, 14 days after the filing of the latter, and a defence filed within 28 days thereafter, failing which judgment may be entered against the defendant without more ado. It is however usually possible to obtain an extension to the 28 day period unless the case is extremely simple.

Statement of claim

The statement of claim is usually in what amounts to two sections. The first expresses the reasons for the claim in legal terms, the second details the particulars of the issues.

Technical content of statement of claim

The statement of claim should, whenever possible, be prepared after at least a reasonable amount of technical investigation has been done and tolerably accurate assessments of cause, value and responsibility for damage arrived at. It is occasionally the case that writs and statements of claim are prepared with no input whatsoever from a suitable technical expert. This course sometimes works but is more often than not the cause of a host of subsequent amendments to the plaintiff's pleadings. (The statement of claim and the subsequent similar and associated documents produced by plaintiff, defendant and third parties are all pleadings.) Such amendments at best are time-consuming, expensive and a source of irritation to all concerned. At worst, the client may be refused permission to amend his case, with consequent difficulties that may prove absolutely critical. Early input from the technical expert is, with rare exceptions, essential.

Responses to a statement of claim

Once received, the statement of claim may produce a number of different reactions from the other side. These will usually include a straightforward defence repudiating some or all of the allegations made, a counterclaim, the joining of third parties or, rarely, an admission of liability.

Further and better particulars of a statement of claim

Assuming however the most likely situation, a straightforward denial of all liability, the defence will be submitted as a pleading, accompanied or followed by a request for details of the plaintiff's allegations, known as 'further and better particulars of the statement of claim'. As the bare minimum of information may have gone into the particulars of the claim this is not altogether surprising. Equally to be expected is a plaintiff's team's subsequent request to the other side for further and better particulars of the defence.

Sometimes requests for such particulars from both sides are almost frivolous in content. They either ask for information which it is impossible for anyone to provide or go to lengths of detail quite unnecessary in the context of the specific allegation about which they are made. Even when of such a nature they can be, at certain stages of the run up to trial, a valuable way of either gaining much needed time before responding to some other request for action, or a means of testing the strength of the case of the other party. It would be churlish to suggest that sometimes they are made purely as an irritant. In the great majority of instances they are a genuine and essential requirement in the assessment of the pleading in question.

Requests for further and better particulars are just that, requests. The party receiving them is under no obligation to reply unless an order to do so is obtained from the court. Such an order may not be granted if the requests are not considered necessary to enable the party making the request to know the case which he has to meet.

An additional requirement to further and better particulars is sometimes called for in the form of interrogatories. These take the form of written questions from one side to another dealing with matters for which there appears to be inadequate documentary evidence. They have to be answered in the form of sworn statements, and may only be obtained with the permission of the court.

The official referee's schedule

This document (commonly called the Scott schedule) is invariably used as a means of presenting all sides of a case in summary form but with the maximum clarity. This, invented, or at least called for, by a judge of earlier times sets out in detail and in schedule format item by item of the technical issues (usually referenced back to the paragraphs in the statement of claim). Typically, details are given of each breach of contract or tort complained of, the damage, the alleged cause, those considered responsible (with reasons for the beliefs) the remedial work needed and its value. Also against each item is the response of each of the defendants to the claim in question with a final column left free for the official referee to make his own notes.

Detailed formats vary, but the above are the principal types of information given. The average Scott schedule tends to be a somewhat unwieldy document, but it is difficult to devise a less cumbersome way of setting down the information, bearing in mind that it is used a great deal in court and has to be available to everyone as a written document rather than something called up on a computer screen. An example of a typical Scott schedule is included in the Appendices, as are examples of pleadings. Unwieldy as it may be in the physical sense, the schedule is a most valuable tool for setting out in a logical way the essentials of the dispute in terms of damage, cause, allegations of responsibility and the defence.

Scott schedules are normally prepared by the plaintiff's counsel, but there will usually be input from the technical side of the team before the document is finalised. Sometimes all the required information can be accommodated within the actual schedule, but where there is considerable repetition it may be cross referenced to separate schedules containing detailed records of findings. This happens particularly, for example, in the case of large buildings or groups of buildings such as local authority housing, where similar defects with similar causes may occur many times, even hundreds, and it would make the document very clumsy to use if there were page after page of identical entries when one, cross referenced to other documents containing the records of the evidence, would suffice.

Timing of the Scott schedule

The timing of the completion and presentation of the schedule to other parties is important. It is another activity that should be undertaken as early as practicable. As with much specially prepared documentation however, the earlier it is completed and available to all parties to the dispute, the greater the likelihood of amendments having to be made later. However, usually the Scott schedule cannot sensibly be prepared in toto until the expert has completed his technical investigations. It is also less likely to need amendment if it is finalised after disclosure by all parties of the documents in their possession and examination of those documents by the plaintiff's legal and technical advisers. The exception is, of course, documents privileged from disclosure.

The timing of discovery, exchange of reports, etc. and orders for directions

At some point there will be orders for directions by the court which set out a timetable for the future progress of the action including, among other things, the latest dates for exchange of experts' reports, discovery (disclosure of all relevant and non-privileged documents to all parties) and a meeting of experts of like disciplines. This latter exercise is called for with the hope that the number of issues to be contested in court can be reduced by agreement among the experts of as many facts and opinions as possible. These meetings will be discussed in greater depth later.

Note that the order in which discovery, exchange of reports and meetings of experts occur may vary. The variations are understandable, as the individual circumstances of a case may affect the logical sequence of events. They do however affect the work of the lawyers and experts and, sometimes, the input of the plaintiff and defendant as well.

By the time that preparations for litigation have been embarked upon the plaintiff's expert will have gone a long way towards preparing his report in its final form. Nevertheless, he and the other experts cannot produce complete reports until after discovery. Thus, if the date for exchange of reports is prior to discovery, all reports will inevitably be qualified to permit amendment or the preparation of supplementary reports later if needed in the light of the information subsequently revealed. Equally, if reports are exchanged prior to the meetings of experts, amendments are highly probable if any matters of substance are agreed. On the other hand, if reports are exchanged after the meeting of experts, there will be other difficulties in connection with the meeting. One will be a reluctance on the part of anybody to discuss anything. Certainly such a course is likely to produce less agreement among the experts rather than more. Perhaps the best basis for an agenda for a meeting of experts is a complete Scott schedule.

From the expert's viewpoint, possibly the ideal sequence of events is discovery, followed by finalisation of Scott schedule, meetings of experts, finalisation of reports and exchange of reports. This may not suit the legal situation however, and in any event, examination of other experts' reports invariably requires comment by each expert on the reports of the others in the form of a supplementary document, although not necessarily for exchange.

Site visits

It is impossible to be definitive about the timing of site investigations and other site visits in relation to other events. Each case has to be treated on its merits, but the principal desiderata related to site inspections by all parties have been covered in Chapter 4. In many cases it will not be necessary for all experts to attend site inspections. Laboratories for example may be given samples of materials and provide expert reports related to their work of analysis, and subsequently give expert evidence. Provided the materials on which they work have been properly attested in relation to the place, date and other circumstances of acquisition, there may be no need for the scientist providing the report to visit the site. Another example is where perhaps a non architect expert has prepared a report covering the incidence, cause and value of damage. If he opines that the original architect was in some way partly or wholly responsible an architect expert may be needed. The architect's report and evidence may well be confined entirely to a comparison of the actual performance of the architect with what he should have done. Very often such reports can be prepared entirely from an examination of documents, although sometimes experts may wish to visit a site just to get the 'feel' of the site and buildings, including orientation, and surrounding buildings or other environment.

More than one expert, 'divide and rule'

If two experts are employed in the sort of situation outlined above there can be problems if they belong to similar related disciplines. Suppose for example that the bulk of the expert evidence has been prepared by a building surveyor and an architect expert has been appointed solely for the reasons given in the previous paragraph. If there were to be any difference of opinion between them whatsoever on technical issues with which both should be familiar there is always the possibility of those differences being exploited by counsel for the opposing side. Even when there appears to be no conflict between the evidence of the two experts it would not be beyond the skill of most counsel to try and suggest it.

How many experts?

Normally, each party to the dispute will have only one expert of each discipline as far as opinions are concerned. It is however not unusual for evidence of fact to be given by a fellow professional to the principal expert. Although this would not be considered as expert evidence in the legal sense, in cross-examination there may well be attempts to obtain opinions from such a witness which could possibly conflict with those of the expert evidence which follows. The type of expert and the total number will be fixed by the court following representations by the lawyers regarding the number of experts they need and why. If one party is permitted to employ an expert to cover a particular specialised subject, the others must be given the same opportunity. In simple cases there may be just one expert employed by each side, while in complex situations there could be as many as four or more. Assume a case involving a plaintiff, four defendants and two or more third parties, each employing four experts, and the prospects in terms of site inspections, meetings of experts, the giving of evidence and costs become absolutely terrifying. Fortunately such extremes are rare, the peripheral experts may not appear in court at all (written statements may well be adequate for some purposes) and the work of secondary experts is mostly fairly limited in scope.

Meetings of experts

Meetings of experts are ordered by the court. The intention is usually to 'agree facts and narrow the issues' in order to save time in court. Such meetings are usually between 'experts of like disciplines', in other words, separate meetings for architect experts, engineer experts, etc. The meetings are on a without prejudice basis, meaning that anything discussed will not prejudice the legal position of any party. The separate meetings for different disciplines can save a considerable amount of time. Often the secondary experts will be able to deal with their aspects of evidence quite briefly. Meetings of the principal experts however can be protracted affairs, frequently with very little of real substance having been agreed at the end of them. A few years ago an all-day meeting attended by 16 experts agreed only that 'Should remedial work be found to be necessary, it will attract VAT at the rate current at the time'. Subsequently the experts discovered that the plaintiffs were a registered charity not subject to VAT, so even that one item of agreement proved to be wrong!

All the meetings are normally convened and hosted by the plaintiff's expert, at a time and venue as far as possible convenient to the others attending. Usually, an agenda is prepared by the host and circulated in advance of the meeting to give the others an opportunity to add items if they so wish. The meetings are not attended by any of the legal advisers. Common items of agreement are matters of fact, 'figures as figures' for example, but without agreeing that the figures are justified, and agreement on methods of identification of buildings, areas where symptoms occur, terminology to be used, etc. Though there is still sometimes some disagreement over what are facts!

Predictably, there will sometimes be agreement between some of the experts on some matters of opinion which result from a common standpoint. For example, experts representing the designer and the contractor may more readily agree that failures have been caused due to poor maintenance by the owner than they will that they have been caused by poor workmanship. With or without complete objectivity by the experts, it is not difficult to see that opinions related to the real grey areas may well be coloured by loyalty to one's client, even if they should not be! In such circumstances the experts can only agree to differ, and the range of opinions may be recorded in the notes of the meeting. Such notes are sometimes typed at the end of the meeting before it breaks up, and signed as a true record by all present. An alternative procedure is for the host expert, who would also chair the meeting, to send out as soon as possible his interpretation of what was agreed and send copies to the others for comment and subsequent ratification when the comments accepted by all as valid are agreed. Items not agreed are also recorded as such, often on the instructions of the official referee. The notes of the meetings are passed back to the expert's instructing solicitor, and such items as are agreed notified to the court. Examples of the latter would be agreeing on a common numbering system for, say, allegedly defective windows, where at an earlier stage each expert had used his own method of numbering.

Agreement of facts

Sometimes specialist experts do agree on some aspects of a case which may save a considerable amount of time. For example, quantity surveyor experts may be able to agree the value of original or remedial work even if the quality of the former or necessity for the latter is left completely out of their brief for others to discuss. Accountants may reach similar agreement regarding, perhaps, consequential damage such as loss of rent. Hence, agreeing 'figures as figures'.

Supplementary reports

Apart from the reason already given, different sequences of events may create a need for supplementary reports dealing with issues as they come up. If exchange of reports takes place prior to discovery there could be a demand for considerable additional input in the form of either amendments to the main report or the production of a supplementary. As with so much in litigation, it is difficult to lay down hard and fast rules. It is necessary to be aware of the possibilities and then treat each occurrence on its merits. The reports of the principal experts may be exchanged before those of the 'specialist' experts, and supplementary reports may be needed to comment on the latter.

The judge's visit

It is the rule rather than the exception for official referees to want to see for themselves at first hand as much as possible of the damage and symptoms of damage. Sometimes by the time legal matters have progressed far enough for a visit to be arranged, even for it to be known who is to try the case, remedial works are well under way or even completed. Despite such circumstances a visit can still be a very valuable means of giving the judge an insight into the nature and extent of the problems. Also, it will give him a picture of the building(s) in question. Both may be of great value when drawings and photographs, etc., are under discussion in court.

The most usual formula for such a visit is for the party to include, in addition to the judge, the principal experts for each side, sometimes more than one from each, and one of the counsel representing each party. (Large and complicated cases usually involve both senior and junior counsel by the time trial is approaching.) It is unusual for instructing solicitors to be represented. Because the majority of buildings are in use during the whole legal process, it is likely that there will also be a representative of the building owner in attendance, if only to assist with access, etc. During the visit the plaintiff's expert will usually make the detailed arrangements for the route to be followed, but this may be varied at the request of other experts as well, of course, as the judge himself.

All present must resist the temptation to give opinions, and restrict themselves to drawing the attention of the judge to any factual matters they feel relevant, and also to answer any questions put to them by the judge. The role of the barristers present is largely to ensure that neither the experts nor owner's representative offer comments which go beyond the stating of facts. It is, however, not unknown for counsel themselves to be questioned by the judge on some aspect of the case.

Settlement negotiations and payments into court

Two other major happenings which commonly occur either before or after commencement of trial are settlement negotiations and payments into court. These are related, and where they are considered at all, which is perhaps the rule rather than the exception, are very much the subject of discussion by all the parties on each side of the dispute. Despite the desirability of reaching a settlement as early as practicable, it is likely that proceedings will have moved far along the road to trial before settlement begins to look like more than a pious hope.

Attempts to settle

It is difficult to generalise about settlement negotiations but there are some matters which can be regarded almost as 'standard', although they will not all occur in every case. Occasionally a move towards settlement is prompted by illness or some similar circumstance. However, in the vast majority of cases the prime factor is money. In the case of the plaintiff making a decision, he may well be influenced by the likely assets of the defendants.

It is probable that initially each side will be looking to establish as far as possible the strength of the hands of the others. This is an essential measure to assist in the making of the 'fight or settle' decision. The making of a decision is, however, seldom clear-cut. Rather than a straightforward choice, it will more probably be questions of when to make an offer, how much should it be and what to do if it is not accepted. Most lawyers would perhaps not be too angry at the suggestion that on occasions during the period after their initial involvement the overall activities of the parties to the dispute resemble not so much the majesty of the law as a skilful but protracted game of poker. There is an added complication. Each hand in the poker game is being held not by one player but by several. They are the central protagonists on each side plus their solicitors, their counsel, and, for good measure, a degree of consultation, usually via the solicitor, with their technical expert. The last of these may need to consult the specialist experts before venturing an opinion on the strength of the case in detail, while the first may have to get agreement at every step from co-directors, financial backers and insurers.

It will be seen that settlement negotiations between two parties are likely to be protracted, and if several parties are involved, extremely difficult and even more lengthy. If there are several parties involved, simultaneous settlement between them all is more likely than between two or more in the first place, followed by either settlement between the remaining parties or, of course, trial.

Settlement between only two of several parties

When in fact one defendant of a number does settle with the plaintiff, it is probable that the settling defendant will demand, before doing so, an indemnity from the claimant against being joined as a third party or sued by one of the other defendants. This indemnity would normally be to cover both damages and costs that may be awarded against the erstwhile defendant in any subsequent action. Most claimants are reluctant to give such an indemnity, which may involve them in considerable extra expense, and in consequence, settlement between plaintiff and one only of two or more defendants is rare.

Because settlement negotiations may take place right up to and beyond the start of trial it is essential for each of the parties and their advisers to try to anticipate likely events related to possible offers of settlement. The possibilities and the permutations will vary of course from case to case, and all the people involved must at all times try and anticipate the likely development of events. The analogy of the poker game should perhaps include a concurrent series of games of chess, with, again, several players involved in each game.

Payments into court

There is one device in common use to reduce the imponderables, namely, the payment into court. Such payments may be made by defendants if they think that they are likely to be found to some extent responsible for the plaintiff's problems, but to a lesser extent than the amount actually claimed against them. A sum which they consider might be accepted by the plaintiff as the limit of what he would actually be awarded is paid to the court where it is held in suspense.

The judge who is to hear the trial is kept in ignorance of the fact that a 'payment in' has been made. The plaintiff, who through his solicitors is advised of the sum, has either to accept or reject the amount. If he accepts, he receives that sum in settlement, plus payment of his costs as assessed by the court up to the date of acceptance. If he does not accept, and the judgment is equal to or less than the amount of the payment in, he receives the amount awarded and assessed costs up to the date of the payment in, but is responsible for payment of the defendant's cost from date of payment in up to the time of the judgment. If judgment is for more than the payment in, then the full amount of the judgment is due from defendant to plaintiff, with costs to be paid as ordered by the court. The system is a potent way of encouraging settlements, provided the amount of the payment in is pitched correctly.

A payment in is also a method of testing the strengths of both the plaintiff's case and his resolve. The initial payment in may be increased at any stage, but obviously, early payment in is virtually an admission of some liability unless of an amount reflecting the commercially based decision of an assessment of how much the defendant is bound to lose anyway in non-recoverable costs.

A defendant making a payment into court has to try and avoid two dangers. If he makes too high a payment, the payment itself will be accepted and the plaintiff will have achieved a settlement upon terms more generous to him than he would have been prepared to accept in negotiations. On the other hand, too low a payment will give the defendant insufficient protection in the litigation. The outline above is, like most of the discussion of legal matters in this book, very much a simplification of what can be a very complicated exercise, and makes no mention of the equally intricate situations that may arise regarding 'payments in' in the event of a case going to appeal.

The pros and cons of settlement

The whole question of settlement, including the ploy of payments into court, is very much bound up not only with the basic claim of the plaintiff and the anticipated judgment in financial terms, but also with the costs involved before and during a hearing. Both plaintiff and defendant as well as any further defendants and/or third parties need to weigh up very carefully three distinct aspects of a dispute:

(a) What is the total amount of the claim excluding fees?
(b) What proportion of the claim is likely to be awarded against each of the defendants? (It may in total add up to considerably less than the amount of the claim.)
(c) What are the likely actual costs that will have to be borne by each of the parties in the event of the case proceeding to judgment and what proportions of those costs are likely to be recoverable from other parties?

The very significant sums that will be expended for photocopying of documents, copies of photographs, travelling, document transmission, etc., will be greatly exceeded by professional fees. These seem to escalate rapidly a few months before trial. Once it has started they may escalate even more rapidly in comparison.

If, as is probable, many of the participants are located some distance from the court there will be considerable hotel costs in addition to travelling, as well as long distance telephone calls and a host of others which mount up at a surprising speed. In addition, fees, expenses and disbursements will attract value added tax, which for some litigants may be partly or wholly irrecoverable.

Each individual situation has to be decided on its merits, but it is highly advisable for all such expenses to be at least estimated and added to estimated fees and related to the amount of the claim before decisions are made about the advisability or otherwise of proceeding further.

To sum up, depending on circumstances, a dispute may be settled at some indeterminate time before the commencement of trial, after the trial has started or may not settle at all. Chapter 11 deals with the second and third of these possibilities, but it may be some encouragement to know that the majority of cases never reach court, and of those that do, many settle shortly after commencement of trial.

Chapter 11

Events in court

Once the date of the hearing has been set, its length estimated and, in the case of litigation, it has not been postponed due to double booking (as the majority of cases settle before trial, dates are frequently double or treble booked) the procedures of trial will commence. The lawyers will liaise with the court to ensure all is ready for the start, and the bundles of trial documents, already prepared and paginated (sequentially numbered), will be taken to court and arranged ready for use. Normally, there will be a set for the official referee or other judge, before whom the hearing is to take place, a set for the witness box and sets for counsel and instructing solicitor acting for each party. In addition to documents and photographs, there may also be samples of materials and/or models, etc.

The bundles of documents will include copies of the pleadings, all relevant correspondence, records of meetings, drawings, etc., that have been used in the preparation of the pleadings. The trial bundles will also include copies of all the experts' reports.

With technical disputes of the type under consideration there is no jury. The order in which witnesses are to be called will have been settled, and the witnesses given, in addition to the venue, some indication of when they are likely to be required in court. As a rule, the case is opened by counsel for the plaintiff who will outline the circumstances in terms of facts, allegations and supporting evidence, sometimes in relatively general terms.

Senior and junior counsel

Depending on the complexity of the case, each party may have as advocates to represent him one or more counsel, and where more than one is necessary there will invariably be a leader and a junior. The former, normally but not necessarily a QC, will usually make the opening speech on behalf of his client and conduct many of the examinations and cross-examinations of witnesses. The junior acts in a supporting role, and will probably do some at least of the examinations and cross-examinations.

Order of calling witnesses

The order in which the witnesses give their evidence is at the discretion of counsel, who may elect to call witnesses of fact before his expert(s) or the other way round. Sometimes some of the witnesses of fact may be before and some after the experts, particularly if there are logistic difficulties for the witnesses themselves in getting to court at a particular time or on a particular day. Essentially however, counsel will try to build up his client's case as a logical progression whenever it is both possible and useful to do so.

Once counsel has outlined his client's case in his opening he will start to examine witnesses, unless it has been arranged that the judge should first visit the site if he has not already done so. If a visit is still required it may not necessarily take place immediately after the opening, but may be at some other suitable time.

Witnesses of fact

Prior to the commencement of their giving evidence, in fact mostly before the start of the trial, witnesses of fact will have written proofs of evidence which will have been made available to the court and all the legal advisers.

The actual examination of witnesses, whether for plaintiff or defendant, follows a regular pattern. The witness is called, enters the box and, depending on religious persuasion, swears an oath to tell the truth, using the appropriate scripture (New or old Testament, Koran, etc.) or, if no scripture is appropriate, the witness may affirm his pledge to tell the truth. He will then be examined, by a series of questions, regarding his own relevant involvement in and knowledge of the case, probably by confirmation of his written and sworn proof of evidence. (The expert's proof of evidence is frequently just his report.) He may be permitted to refer to the documents provided for that purpose, in fact he will normally be invited by counsel to look at particular documents and comment on various aspects of them in response to questions. He may, during the course of giving evidence, be asked directly by the judge to clarify or elaborate on his answers. When he has answered all the questions put to him by counsel he will, in the majority of instances, be cross-examined by counsel for the other side.

If there are several defendants and/or third parties he may well be cross-examined by counsel for each of them. After cross-examination he may be re-examined by his own counsel. (More accurately, the counsel for the party who has called him as a witness.)

During the examination of witnesses there is no automatic bar on the presence in court of witnesses still to be called. One rule that must be observed is that if the court adjourns, for whatever reason, a witness who at the adjournment is still on oath may not discuss the case with anyone.

People who are to appear in the witness box for the first time tend to regard the prospect as daunting. Many who have often done so are prepared to admit that it remains daunting. The only advice it is possible to offer to witnesses about to give evidence for the first time, or the twenty-first come to that, is to observe a few basic rules, the principal ones being these:

(1) Speak up, speak clearly, towards the judge. (Your solicitor will tell you beforehand how to address him if you have to respond to his questions.)

(2) Remember that although the proceedings are these days usually tape recorded, the judge will be making his own notes, give him time to do so.

(3) Answer questions as succinctly as possible, but don't be stampeded into answering yes or no to questions of the 'have you stopped beating your wife' type.

(4) Don't be afraid to say that you do not know the answer to a question, have forgotten, want more time to think or need reference to some note or other document. (Better any of those than guessing wrongly!).

Expert witnesses

The examination of expert witnesses follows a similar pattern to the others. That is, examination in chief, cross-examination and re-examination. They may be questioned more often than other witnesses by the judge, usually to assist in the clarification of technical issues that may be somewhat difficult for comparative laymen (laymen in the sense of building design and construction that is) to appreciate fully. This is almost inevitable to some extent, no matter how carefully expert's reports are written, and how carefully technical terms are explained. The experts are also likely to be taken through their reports in considerable detail, both during examination in chief and cross-examination.

Usually, a blackboard or similar provision for making sketches is provided adjacent to the witness stand. However clear the drawings and diagrams in the reports of the various experts, there are frequently points that arise that require 'instant' drawings to assist clarification.

The expert's role out of the witness box

Experts are also frequently needed in court for considerable periods out of the witness box. Because of their extensive knowledge of the technical aspects of a case they have an additional role to that of actually giving evidence. During the presence of other witnesses in the box, particularly the opposition experts, they are often needed to make a very quick analysis of the answers given by those witnesses. This is in order to advise their own counsel of the import of the answers given and perhaps to suggest questions which the witness should be asked during cross-examination. This will normally be achieved by note taking while the court is sitting and discussion during adjournments. Sometimes it may be useful to alert counsel immediately to some technical point while he is cross examining. This is done by the expert passing a note to the instructing solicitor, usually sitting immediately in front of his expert and behind counsel. The solicitor will decide on the wisdom or otherwise of passing the information immediately to counsel.

The desirability of having the expert in court for a large part of the proceedings has to be weighed against the cost involved. All the time that the expert is unable to be doing other work will be charged to his client and ultimately to whoever has to pay costs. Nevertheless, there can be dangers in not having the expert hearing other witnesses' evidence. In a recent case the evidence of the plaintiff's expert was heard before that of witnesses of fact and the former was not in court when one of the witnesses described, erroneously, an investigation carried out on site by the expert, with unfortunate consequences for the plaintiff. Had the expert either heard or been given a transcript of the evidence of the factual witness before going in the box himself an attempt could have been made to remedy the damage. The culprit was however, really, the expert for not fully describing the tests he did on site which would have avoided any possibility of ambiguity and also the need for the witness to be questioned on the subject. It should have been detailed correctly in the expert's report.

The plaintiff's experts

It is re-emphasised that it is most important for experts to be completely objective and state their opinions fully, accurately and honestly. In theory therefore there should be no need for more than one set of experts acting to assist the court rather than acting for their own client, be he plaintiff or defendant. However, no matter how genuine the opinion of one man it may differ from that of another, and the expression of different opinions and the arguments on which they are based are of value in assisting the court to make a judgment. Sometimes the process may be assisted, to the discomfiture of the relevant team, when two experts on the same side have different opinions. When this happens, opposing counsel will not be slow to try and drive a wedge between the witnesses involved, and attempts by a witness to repair the damage may well make things worse. It is usually accepted by the Court that a witness may reasonably leave to his fellow experts matters which are on the periphery of his experience. A refusal to hazard an opinion on subjects outside one's experience is difficult to argue against.

However, there are occasions when two witnesses may each be expected to have detailed knowledge of a particular issue, and it is essential that such differences are brought out not merely before trial but before finalisation of reports, as if they are important they could affect the wisdom of fighting the case at all.

Apart from two experts on the same side, there is another aspect of giving evidence that the plaintiff's team must bear in mind. In one-day cricket it is sometimes considered an advantage to field first, as by batting last a team knows its commitment in advance. To some extent the plaintiff's expert is, in a similar way to the side batting first, at a slight disadvantage in that, no matter how comprehensive his report, there will almost always be factual matters brought out in cross examination that the opposition will have time to think about before their own witnesses give evidence. Once his own evidence is finished, it will be incumbent upon his client's counsel to deal with such events, probably when cross-examining the defence witnesses, expert or otherwise.

Such unforeseen matters must be genuinely so. The courts do not permit deliberate 'surprises' in the introduction of new evidence. Where new evidence is important enough for its introduction to be justified, the opposing side(s) must be given time to consider the implications and then have the opportunity to deal with it.

Defendant's and third parties' experts

Batting last can, as suggested, be a positive advantage. Also it is sometimes easier to demolish a case than to make the case in the first place. On the other hand, the majority of actions are not brought without a belief by the plaintiff and his team that they have good grounds for fighting. Nevertheless, the defendant's experts, despite being just as objective as the others, may, by the mere fact of having an opportunity to listen to the evidence and opinions being elaborated on, be able to take advantage of not being first to testify. For example, the initially heard experts may state that they have relied upon particular sources of information. Research may demonstrate that the information used was out of date or incomplete. Neither the original statement of what was used nor the results of the subsequent research would necessarily be classed as 'surprises'. .

Apart from this batting last advantage, the role of the defendant's expert is very similar to that of the plaintiff's. There is the same requirement to listen to the examination and cross examination of other witnesses in order to feed counsel with technical information if necessary.

Checking facts elsewhere than court

Sometimes, points will arise that cannot be answered immediately and the expert may be asked to produce the answer for the next sitting of the court, maybe after lunch, the following morning or after a week-end break. Sometimes the timetables set are extremely difficult to meet, but they must be met unless it is clearly impossible. It is not always the expert who is involved, or perhaps not the expert on his own. Such requests can mean working unsocial hours under considerable pressure. The better the report, the better the case is prepared, the less likelihood of such situations arising.

The defences

Once the plaintiff's counsel has completed his examination of all his witnesses, and they have in turn been cross-examined and re-examined, counsel will advise the judge that he has no further witnesses and the defendant's and third party counsels will in turn open their respective cases and call their witnesses. These will be examined and cross-examined in a similar manner to those of the plaintiff. Once all the witnesses have been heard, the senior counsel for each party will have an opportunity to make a closing submission to the court, as well as, of course, counsel for the plaintiff. These may be oral, written or both.

Betterment

The question of betterment will almost invariably come up during trial. It is often of considerable importance, with the defendants endeavouring to demonstrate that much of the remedial work carried out or intended to be carried out by the plaintiff is in fact unnecessary as far as putting the building into good order is concerned. Arguments for and against will usually cover both the need for the work itself and also its cost, though the latter of course will apply to all remedial work including that which is accepted as being essential. While any work not forming an essential repair or replacement may be betterment, there are often grey areas relating to work which create improvements in some respects but which at the same time are the most economical method of making the building whole. Also taken into consideration will be the amount of useful life that replaced components have already had related to the life of the building as a whole.

Submissions by all parties will invariably involve not only the principal experts on each side but also quantity surveyors and, where the remedial works have not been designed by the principal expert, whoever has in fact been responsible for the design. There is little guidance that can be given on this subject, apart from saying that each case must be treated on its merits. This should preferably be done as part of the plaintiff's expert's report, and should demonstrate clearly what is betterment and why it has been done. In addition, if grey areas are defined as such, and the need for the work demonstrated clearly, much time and possible injustice may be saved.

Judgment

Once the trial has been fully heard, the parties will await judgment. As far as the official referee's courts are concerned, the pressure of work is such that it may take a considerable time for the judgment to be ready. When it is, the lawyers for each party are advised of the date and time at which it will be given. Subsequent to the reading of the judgment it is made available to the various parties in written form, and eventually may be published in the *Law Reports*.

Orders for damages

Usually, part of the judgment will be to award damages relating to the parties to the dispute, assuming that the plaintiff is successful. Sometimes however the judgment stipulates the nature and degree of liability in the form of negligence or breach of contract, if any, of each party, but leaves the amount of damages to be agreed between them. This procedure may be facilitated by the parties having previously agreed 'figures as figures' without having admitted or agreed liability.

Such a procedure would be covered by a requirement that the contenders would have to return to the court for a ruling in the event of voluntary agreement not being reached.

Orders for costs

Judgments also cover the question of apportioning costs. While it is usual for the losers of a case to be expected to pay between them, in varying proportions, at least part of the winner's costs as well as their own, costs are frequently disputed by the parties with regard to what is and what is not reasonable. Where this happens the costs are put to an official of the court for a ruling. The official is known as the taxing master, and his ruling fixes the taxed costs payable by each party, which usually amount to between 60% and 80% of the total costs of the successful party notwithstanding that, it should be appreciated that all matters relating to costs, including, incidentally, payments into Court, are at the overriding discretion of the Court.

Appeals

If any of the parties feel at the end of a hearing that the court has in some way been in error they may wish to appeal. If the appeal is over a point of law, the right to appeal must be granted. If the grounds are on questions of fact, leave to appeal may by rejected by the official referee. Such rejection may itself be referred to and overturned by the Court of Appeal. There may also in some circumstances be special leave to appeal as to costs. Should leave to appeal be granted, the Appeal Court will go through the documents in the case, the transcripts of evidence and the judgment. They do not hear witnesses, and reach a decision purely on the basis of the written word plus representations by counsel giving reasons for dissatisfaction. The appeal decision may in rare instances be referred to the House of Lords, who are the final arbiters.

Chapter 12

Conclusions

Summing up the process of attempting to recover the costs resulting from building defects, or, alternatively to try and refute the allegations of responsibility, remember the initial triggers. Perhaps water where it should not be, materials falling off the building, alarming cracks appearing in walls or ceilings, or some similar indication that all is not well.

The initial reaction, depending on the degree of emergency, is to effect temporary repairs, followed by permanent repairs, invariably after some attempt to assess the likelihood of them being paid for by someone else. If the someone else is not the owner's insurer, then who is it, and what chances are there of success? How does one get expert advice, both legal and technical, and what is it going to cost? Will the cost of that advice be recoverable, and what other consequential costs may be legitimately claimed?

In attempting to answer those questions, and the questions that invariably follow, the previous eleven chapters tried to achieve two primary objects. The first to give an outline of the procedure to be followed after symptoms of failure or defect have been found in a building. The second to be as realistic as possible about these events and, if anything, to err on the side of pessimism when describing some of the things that matter to the person suffering the defects and the persons alleged to have been responsible for them.

In going through events from first symptoms to the final judgment of the court, the pessimism has not been overdone. What is slightly happier though is that while most disputes get as far as the serving of a writ, the great majority are settled before they reach the courtroom.

That does not alter the need to be aware, whatever the nature of the dispute, that it may finish up as a full-blown litigation, be it in the High Court, at County Court level or perhaps as arbitration. There are, therefore, considerable advantages in knowing beforehand to some extent at least the procedures involved and what they are likely to mean in terms of time and cost. It may be argued that the more all sides know about what they are likely to be in for if a case does go to trial, the less likely it is to do so!

Also, the fewer the number of parties there are, the greater the likelihood of settlement being sooner rather than later. As well, of course, if there are only two parties in contention, there is the possibility of arbitration as an answer if agreement cannot be reached by negotiation.

Whatever happens in any particular dispute, the people immediately involved, the litigants, unless they manage to settle their differences without help from others, will each be a member of a team, and the more each member knows about the work of the other members, the more smoothly, quickly and cheaply will matters progress. Each member of the team other than the litigant must remember that with every further move they are committing someone else's resources. Partly for that reason the reader is reminded of the need for all the advisers to tell their client when his case is a poor one. This can only be done if the client, be he potential plaintiff or potential defendant, is comprehensive and utterly frank with his briefing to those advisers, and with the background information that he gives them. Once he has been so it is up to his legal and technical experts to play their part by being completely objective with the advice that they give and with the representations that they may ultimately make to a court. The essential thing for all those involved to remember is that the operative word is indeed team. Once that is realised it is also easy to realise that not only do all have a part to play, but they must recognise to the full the part played by each of the others.

To play their part correctly each must be kept informed as early and as fully as possible of any actions by others (of their own team or that of the opposition) that may impinge upon their own activities. This can only be done if each does indeed know how the others operate, what their work entails, how busy they are, and in consequence, how much time they need to undertake any particular activity or series of activities.

The preceding chapters have attempted to give some idea of the events that occur as part of the resolution of a dispute, the relative importance of those events, and the sequence in which they occur or the sequences in which they might occur. Many matters, particularly legal aspects, have deliberately been dealt with in only general terms. There are some which will very probably crop up that have not been covered at all. This is almost inevitable because, even though there may be similarities, no two situations are identical, and, in addition, case law is continually being added to and creating precedents that make it impossible to foresee the situation that may exist in legal terms at the start of the preparation for an action.

There remains in consequence at least one point of importance which needs some elaboration. The crucial need to obtain appropriate advice at an early stage has been emphasised. Touched upon, but needing greater emphasis, is the importance of each case being treated strictly on its merits at the time it arises, and, indeed, the merits being reconsidered as events unfold. This applies both to technical and legal aspects. Publication of research, for example, may need a completely unexpected revision of opinion at any stage, while a judgment or outcome of appeal in a similar case to the one under consideration may need a complete rethink of future legal action. These happenings are fairly rare, but nevertheless the consequences if they do occur can be dramatic. Technical and legal advisers are on guard against the possibilities of such situations arising, but it is necessary for litigants to be aware that at any stage between initial defect and appearance in court the prognosis for the outcome of an action may change.

Overall, there is no doubt that if disputes can be settled between the parties with the aid of a mixture of goodwill and common sense, and without the aid of lawyers, technical experts and the courts, a great deal of time, frustration and money is likely to be saved. Unfortunately, because there are nearly always two sides or more to any argument the ideal situation is very much the exception. At the other end of the scale, the majority of disputes do not get as far as the court, and many that do still settle without running their full term.

It would be nice to conclude by saying that the earlier a case settles, the more satisfactory in every way for all the protagonists. This may in most instances be true, but regrettably, not all. In the end, the only certain advice for everyone involved is to be constantly objective, treat each situation strictly on its merits, and remember that they are indeed one of a team in which each has a part to play and needs time to play it.

Appendices

The documents that follow are fictitious examples of basic pleadings. They include a statement of claim, a defence, an order of the court, a reply to the defence, request for further and better particulars of the statement of claim, and the further and better particulars themselves. There is also part of the relevant official referee's (Scott) schedule. This last document, included, like the others, to show a typical format, only deals with some of the allegations in the statement of claim. In a real situation it would of course deal with all.

A
Statement of claim

IN THE HIGH COURT OF JUSTICE   1990-ORB-No.66666
QUEEN'S BENCH DIVISION
OFFICIAL REFEREES' BUSINESS
HIS HONOUR JUDGE PERCY PARITY Q.C.

BETWEEN:-

THE BLOGGS SHIPPING COMPANY LIMITED
Plaintiffs

-and-

FLAUNDEN AND CHESHAM (a firm)
Defendants

STATEMENT OF CLAIM

1. The Plaintiffs are the owners of Bloggs House, ("the
    building"), at Bodgit Major in the county of Southshire.

2. At all material times the Defendants were in practice as
    architects.

3. By an agreement is writing dated 4th July 1980 the
    Plaintiffs engaged the Defendants to be their architects
    for the design and inspection of the construction of the
    building.

    The said agreement provided inter alia that:-

    -    by clause 7: that the Defendants' appointment was as
         architects.

- by clause 8: that the Defendants would, as architects, exercise all proper and reasonable skill, care and diligence in conformity with the normal standards of the architect's profession in designing Bloggs House, and in the monitoring and inspection of the construction works.

4. The building suffers from water penetration caused by the Defendants' breach of the contract entered into with the Plaintiffs in that the Defendants did not exercise the skill and care of a reasonably competent architect, in accordance with clause 8 of the contract, in or about the design of the building and in or about the inspection of the construction works.

PARTICULARS

(1) <u>Basement</u>

Much of the basement is situated below a large paved area which abuts the ground floor of the building. The design and construction of the building has not provided adequate means of disposing of surface water falling into the paved area. As a result, water has penetrated the basement by a variety of routes of which the following are the most important:-

(i) through the junction between the outer walls of the main building and the paving because:

(a) The Defendants failed to ensure that the asphalt upstand to the external wall of the main building was of sufficient height.
(b) The Defendants failed to ensure that the upstand was chased into the brickwork.
(c) The Defendants failed to ensure that a flashing was provided over the upstand.
(d) The Defendants failed to provide a waterproof detail at the junction between the paving slabs and the asphalt upstand.

(ii) At the junction between the bottom of the brick-clad concrete columns and the paving because:

> (a)   The Defendants failed to ensure that the top of the asphalt upstand to the columns was of sufficient height.
> (b)   The Defendants failed to ensure that there was a flashing over the upstand.
> (c)   The Defendants failed to ensure that the upstand was chased into the brickwork of the columns.
> (d)   The Defendants failed to ensure that weepholes were provided in the brickwork.
> (e)   The Defendants failed to provide a waterproof joint between the asphalt upstand and the columns of the one hand and the basement roof deck on the other.

(ii) <u>Staircase</u>

> Damp staining has occurred to the reveals of the staircase windows because the Defendants failed to ensure that the vertical damp proof course was properly installed in the sides of the window frames, resulting in the plaster bridging the brick cavity and making contact with the damp outer leaf.

5.   By reason of the matters set out in paragraph 4 above the Plaintiffs have suffered loss and damage.

PARTICULARS

<u>ITEM</u>		<u>COST</u>
(a)	Temporary remedial works to the basement	£27,570.00
(b)	Main remedial works to the basement	£578,420.00

(c)	Work to window reveals	£125,000.00
(d)	Professional fees on the above @ 20%	£146,198.00
(e)	Value Added Tax on the above @ 15%	£131,578.20

Total £1,008,766.20

Full particulars of the work undertaken will be provided in due course in the form of an Official Referee's schedule.

6. The Plaintiffs are entitled to and claim interest pursuant to S35A of the Supreme Court Act 1981 at such rate as the Court shall think fit.

AND THE PLAINTIFFS CLAIM

(1) Damages under paragraph 4.

(2) Interest under the said statute

A. D'VOCATE

Served this 1st day of October 1990 by Messrs. Cholomondley, Cholomondley and Chumleigh of 17, Bagots Rents, Loose Chippings, Southshire, Solicitors for the Plaintiffs.

B
Defence

IN THE HIGH COURT OF JUSTICE   1990-ORB-No.66666
QUEEN'S BENCH DIVISION
OFFICIAL REFEREES' BUSINESS
HIS HONOUR JUDGE PERCY PARITY Q.C.

BETWEEN:-

THE BLOGGS SHIPPING COMPANY LIMITED
Plaintiffs

-and-

FLAUNDEN AND CHESHAM (a firm)
Defendants

_____

DEFENCE
_____

1. The Defendants admit paragraph 1 of the Statement of Claim;

2. The Defendants admit paragraph 2 of the Statement of Claim;

3. The Defendants admit paragraph 3 of the Statement of
    Claim, but will rely on the whole of the said agreement at
    trial for its full terms, true meaning and effect;

4. It is denied that the Defendants were in breach of the
    duties set out in paragraph 3 of the Statement of Claim as
    alleged or at all.

5. It is further denied that any such breach of contract, which is denied, caused the alleged loss or damage.

6. The Defendants deny that the Plaintiffs have suffered loss either as alleged at paragraph 5 of the Statement of Claim or at all.

7. Further, and in the alternative, if, which is denied, the building is defective as alleged or at all, the remedial works carried out were unnecessary and/or constitute betterment and the cost thereof is not recoverable.

8. Further, if, which is not admitted, the Defendants were in breach of contract as alleged or at all, the Plaintiffs' claim is statute barred, any cause of action having accrued by 31st August 1983.

9. Save as hereinbefore expressly admitted, each and every allegation contained in the Statement of Claim is denied as if the same were set out more particularly below and traversed seriatim.

10. In the premises the Defendants deny that the Plaintiffs are entitled to the relief sought or to any relief.

E. SILVERTONGUE

Served this 15th day of October 1990, by Messrs. Dobermann, Pinscher, Rottweiler and Co., of 23, Savage Buildings, Snarl Street, Loose Chippings, Southshire. Solicitors for the Defendants.

C
Order of the court

<pre>
IN THE HIGH COURT OF JUSTICE   1990-ORB-No.66666
QUEEN'S BENCH DIVISION
OFFICIAL REFEREES' BUSINESS
HIS HONOUR JUDGE PERCY PARITY Q.C.
</pre>

BETWEEN:-

THE BLOGGS SHIPPING COMPANY LIMITED
                                        Plaintiffs

-and-

FLAUNDEN AND CHESHAM (a firm)
                                        Defendants

ORDER

UPON HEARING Counsel for the Plaintiffs and the Defendants

IT IS ORDERED THAT:-

1. The Plaintiffs do serve their Reply on or before
    3rd December 1990 if so advised;

2. The Plaintiffs do serve Further and Better Particulars of
    the Statement of Claim on or before 3rd December 1990;

3. The Plaintiffs do particularise their allegations of
    defects in an Official Referee's schedule under the
    following headings; item number; location; defect; cause;
    allegations against the Defendants; reference to original

contract drawings; reference to remedial work drawings; remedial work necessary; cost of remedial work; Defendants' comments; Official Referee's comments. The schedule to be served on or before 3rd January 1991;

4. The Defendants do serve their comments on the Official Referees' Schedule on or before 5th February 1991;

5. The parties do exchange lists of documents which are or have been within the power, possession or control of the parties and which relate to the matters in question in this action, on or before 11th February 1991 with inspection on seven days' notice;

6. The parties do respectively have leave to call expert evidence limited to two Expert Witnesses each;

7. Experts as to liability to meet without prejudice on or before 10th April 1991;

8. Experts as to quantum to meet without prejudice on or before 15th April 1991;

9. Following each meeting of Experts, the Expert for the Plaintiffs to prepare and circulate a statement as to those matters upon which the respective experts are agreed and as to those upon which they are not agreed;

10. Experts' Reports as to Liability and as to quantum be exchanged on or before 31st May 1991;

11. Duly signed statements of witnesses of fact be exchanged on or before 10th June 1991;

12. The Plaintiffs restore the Summons for the Pre-Trial Review of one hour's duration on 14th June 1991 at 10.00am;

13. The trial of the Action be fixed for 1st July 1991 with an estimated duration of ten days, first fixture;

14. There be liberty to apply.

   AND that the costs of and occasioned by this application
   be costs in the cause with a certificate for Counsel.

   Dated this 26th day of November 1990.

D
Reply to a defence

IN THE HIGH COURT OF JUSTICE   1990-ORB-No.66666
QUEEN'S BENCH DIVISION
OFFICIAL REFEREES' BUSINESS
HIS HONOUR JUDGE PERCY PARITY Q.C

BETWEEN:-

THE BLOGGS SHIPPING COMPANY LIMITED
                                    Plaintiffs

-and-

FLAUNDEN AND CHESHAM (a firm)
                                    Defendants

REPLY

----

1. The Plaintiffs join issue with the Defendants on their
    Defence except to the extent that it consists of
    admissions.

2. The agreement between the Plaintiffs and the Defendants
    was executed under seal. Accordingly, the Plaintiffs deny
    that their claim is statute barred as pleaded or at all.

    SERVED this 2nd day of December 1990 by Messrs.
    Cholomondley, Cholomondley and Chumleigh of 17, Bagots
    Rents, Loose Chippings, Southshire, Solicitors for the
    Plaintiffs.

E
Request for further and better particulars of the statement of claim

IN THE HIGH COURT OF JUSTICE   1990-ORB-No.66666
QUEEN'S BENCH DIVISION
OFFICIAL REFEREES' BUSINESS
HIS HONOUR JUDGE PERCY PARITY Q.C.

BETWEEN:-

THE BLOGGS SHIPPING COMPANY LIMITED

Plaintiffs

-and-

FLAUNDEN AND CHESHAM (a firm)

Defendants

REQUEST FOR FURTHER AND
BETTER PARTICULARS OF THE
STATEMENT OF CLAIM

Under paragraph 4

of: "The building suffers from water penetration caused
     by the Defendants' breach of the contract ... in that
     the Defendants did not exercise the skill and care of
     a reasonably competent architect"

Request
Specify exactly what acts or omissions constituted the
Defendants' alleged breach of contract.

Request
Specify precisely by reference to a plan or sketch of the
building each and every location and route of water
penetration.

Request
Specify the date upon which water penetration first
occurred.

                                         E. SILVERTONGUE

Served this 15th day of October 1990 by Messrs. Dobermann,
Pinscher, Rottweiler and Co., of 23, Savage Buildings, Snarl
Street, Loose Chippings, Southshire.   Solicitors for the
Defendants.

F
Further and better particulars

IN THE HIGH COURT OF JUSTICE   1990-ORB-No.66666
QUEEN'S BENCH DIVISION
OFFICIAL REFEREES' BUSINESS
HIS HONOUR JUDGE PERCY PARITY Q.C.

BETWEEN:-

THE BLOGGS SHIPPING COMPANY LIMITED

Plaintiffs

-and-

FLAUNDEN AND CHESHAM (a firm)

Defendants

FURTHER AND BETTER PARTICULARS
OF THE STATEMENT OF CLAIM
PURSUANT TO THE DEFENDANTS'
REQUEST DATED 15th OCTOBER 1990

Under paragraph 4

of: "The building suffers from water penetration caused
by the Defendants' breach of the contract ... in that
the Defendants did not exercise the skill and care of
a reasonably competent architect"

Request
Specify exactly what acts or omissions constituted the
Defendants' alleged breach of contract.

Answer
The Defendants failed to use reasonable skill and
care when performing design and inspection services for
the Plaintiffs pursuant to the agreement between them. In

so breaching the contract the Defendants failed to
exercise the degree of reasonable skill and care which a
reasonably competent architect, in the circumstances,
would have taken. The particular design errors
constituting the breach of contract will be set out more
particularly in the Official Referees' schedule.

Request
Specify precisely by reference to a plan or sketch of the
building each and every location and route of water
penetration.

Answer
Please refer to the sketches served herewith.

Request
Specify the date upon which water penetration first
occurred.

Answer
In or about December 1984.

                              A. D'VOCATE

Served this 3rd day of December 1990 by Messrs.
Cholomondley, Cholomondley and Chumleigh of 17, Bagots
Rents, Loose Chippings, Southshire, Solicitors for the
Plaintiffs.

G
Official referee's (Scott) schedule

IN THE HIGH COURT OF JUSTICE   1990-ORB-No.66666
QUEEN'S BENCH DIVISION
OFFICIAL REFEREES' BUSINESS
HIS HONOUR JUDGE PERCY PARITY Q.C.

BETWEEN:-

THE BLOGGS SHIPPING COMPANY LIMITED

Plaintiffs

-and-

FLAUNDEN AND CHESHAM (a firm)

Defendants

OFFICIAL REFEREES' SCHEDULE

Item	Location	Defect	Cause	Allegations against the Defendants
(1)	(2)	(3)	(4)	(5)
(a) (i)	Junction between the outer walls of the main building and the paving.	Water penetration of the basement at high level.	(1) The asphalt upstand to the external wall of the main building is of insufficient height in contravention of good practice and CP 144.	Failed to ensure that the upstand to the external wall of the main building was of sufficient height in accordance with good practice and CP 144.
			(2) The asphalt upstand is not chased into the brickwork and/or the concrete in accordance with good practice or CP 144.	Failed to ensure that the asphalt was chased into the brickwork and or the concrete in accordance with good practice and CP 144.
			(3) There is no flashing over the upstand.	Failed to ensure that flashing was provided on the upstand.
			(4) The detail at the junction between the paving slabs and the asphalt upstand is not waterproof	Failed to provide a waterproof detail at the junction between the paving slabs and the asphalt upstand.

Reference to original contract drawings	Reference to remedial contract drawings	Remedial work necessary*	Cost of remedial work	Defendants' comments	Official referee's comments
(6)	(7)	(8)	(9)	(10)	(11)
FC/372 REV T	CD/1 REV A	(1) Replace upstand on expanded metal lathing fixed to brickwork.	Reference should be made to the Bills of Quantities.	As is apparent from the fact that the specification for the remedial work substantially follows the original specification it is clear that the faults complained of do not result from any design errors, but are attributable to failures of workmanship.	
FC/373 REV 2	CD/4	(2) and (3) Fix non-ferrous metal flashing over new asphalt upstand 150 mm above finished external level and chase into brick.			
FC/373 REV T	CD/8	(4) Amend detail at paving slabs and asphalt upstand.			
FC/381/REV Y	CD/2				

* *Note:* This is intended as a brief indication of the remedial work necessary to prevent water penetration. It is not a detailed description of the work to be done or a specification and should not be worked to. Reference should be made to the remedial works contract documentation, copies of which are served herewith.

Further Reading

Addleson L. (1982) *Building Failures*. Butterworth Architecture.

Eldridge F. (1976) *Common Defects in Buildings*. HM Stationery Office, London.

Mildred R.H. (1982) *The Expert Witness*. George Godwin.

Parris J. (1983) *Arbitration, Principles and Practice*. BSP Professional Books, Oxford.

Ransom W.H. (1987) *Building Failures: diagnosis & avoidance*. Spons.

Reynolds M.P. & King P.S.D. (1988) *The Expert Witness and his Evidence*. BSP Professional Books, Oxford.

Say E. (1983) *Official Referee's Business*. Sweet & Maxwell.

Index